# TRAVEL WRITING FOR TOURISM AND CITY BRANDING

*Travel Writing for Tourism and City Branding* is an insightful, expert-led book which provides tourism students with a practice-based approach to producing researched literary travel writing on an urban destination, using the writing process as a research tool in itself. The book is scientifically supported with full academic references for researchers.

On a global basis, city councils and destination managers are seeking new ways to commission and sponsor professional content authors as part of place-branding projects for tourism development. Given the increasing prevalence of such content within the tourism industry, this book provides a cohesive overview of literary travel writing, presenting it as an enquiry process that can be applied by writer-researchers to spaces that have value to them. Travel writing is presented as a methodological practice that researchers can learn and apply to their own projects, both in academic settings and in commercial city branding. Examples of literary travel writing are carefully examined throughout and their affects refracted through further work. Enriched with a wealth of case studies, chapters are presented in such a way that readers can take the work as a model for their own projects.

This informative and practical volume will be of great interest to students of tourism marketing, destination marketing, place branding and travel writing, as well as current creators of commercial tourism marketing content.

**Charlie Mansfield** has been a university lecturer since 1995 and taught until January 2022 at the University of Plymouth in tourism management and French, where he was also co-director of the heritage research centre. He completed a major, funded research project for the CNRS with the University of Paris 1 Panthéon-Sorbonne in digital heritage management and was a research academic with the University of Edinburgh from January 2005 until July 2009 where he successfully completed an AHRC-funded research project to digitise medieval literature. He is an independent researcher and travel writer, regularly running summer schools for literary travel writers.

**Jasna Potočnik Topler** has been teaching at the University of Maribor, Slovenia, since 2014. She completed her studies in English language and literature and in journalism, and was awarded her PhD at the University of Ljubljana. Currently, she holds the position of associate professor. Her field of research includes cultural tourism with its subtypes, languages, tourism and media discourse, and communication. She is the author of monographs, scientific articles and conference lectures, and an editorial board member of many journals. She has been engaged in several international projects and in projects with students, the most recent under the EU Erasmus+ Programme KA220-HED is IN-COMM GUIDE that enhances active and inclusive teaching of literacy and communication skills for better employment and sustainable economic growth. As a guest lecturer, she has had cooperation with many European universities.

# TRAVEL WRITING FOR TOURISM AND CITY BRANDING

## Urban Place-Writing Methodologies

*Charlie Mansfield and Jasna Potočnik Topler*

Routledge
Taylor & Francis Group

LONDON AND NEW YORK

Designed cover image: Getty Images. Photograph shows the Chehel Sotoun pavilion of the forty columns in Isfahan.

First published 2023
by Routledge
4 Park Square, Milton Park, Abingdon, Oxon OX14 4RN

and by Routledge
605 Third Avenue, New York, NY 10158

*Routledge is an imprint of the Taylor & Francis Group, an informa business*

*British Library Cataloguing-in-Publication Data*
A catalogue record for this book is available from the British Library

ISBN: 978-1-032-01472-2 (hbk)
ISBN: 978-1-032-01469-2 (pbk)
ISBN: 978-1-003-17878-1 (ebk)

DOI: 10.4324/9781003178781

Typeset in Times New Roman
by codeMantra

# CONTENTS

# ILLUSTRATIONS

## Figures

## Tables

# PREFACE

Travel writing is gaining importance and popularity among academics, researchers and communication experts in the tourism industry. Research postgraduates in tourism management studies and new media professionals are embarking on travel writing projects and yet they lack preparation for their fieldwork. Further, they are often unclear about how the finished text should appear, how it will be published and distributed and how it will be valued. In the two disciplines of tourism studies degrees and creative writing courses, new methodologies remain undeveloped and are not reviewed in the academic journal literature, while increasing numbers of researchers want to use their travel writing as a valid research method. This book addresses that gap.

City councils and destination managers (DMOs) are seeking new ways to commission and sponsor professional content authors as part of place-branding projects for tourism development and for inward investment to towns and cities; this book will aid the professionalisation of this sector of the creative industry. The approach in this text builds an exciting new link between tourism and city branding that will engage new travel writers and bloggers in sensitive and highly valued content creation.

The book proposes that the practice of literary travel writing is an inquiry process that can be applied by writer-researchers to spaces where value is being created. The inquiry process elicits, stores and communicates that value and at the same time provides a discovery instrument for writers to interrogate the place further. This travel writing as inquiry is presented as a methodological practice that researchers can learn and apply to their own research projects in both academic settings and commercial city branding projects. Examples of literary travel writing from research projects by the authors are examined and their affects are refracted through further work. The book also reports on the research findings on urban spaces using this new method, as examples of the effectiveness of this approach. In some cases, these findings are synthesised in narrative discourse to make a contribution to new knowledge in a literary form. It is hoped that future researchers and strategists will judge these synthesised outputs and their impacts on a public readership and on policy makers and, from this, continue the dialogue inherent in this research methodology.

Well-researched, literary travel writing on an urban destination provides a valuable resource for content providers to use on social media channels. The narratives collected also contribute to the brand story of a city and tourism destination. This storytelling reaches potential visitors through narrative knowing, which is qualitatively different from, and yet supplements, technical and scientific knowledge. Through the innovative journaling system presented in the book, the structures on which stories can be built are explained in detail alongside a new framework for sharing creative work dynamically with sponsors or commissioning bodies.

Throughout the whole text, the chapters and case studies are presented in such a way that readers can take the work as a model for their own research projects and writing. This includes both research students and destination managers but also lecturers setting work for and assessing undergraduate and masters groups. To support this, a very thorough connection is made with current research theory from the disciplines that make up tourism and heritage knowledge. References to academic articles are made throughout the text, and a complete list of references is made available at the end of the book.

# INTRODUCTION

## Who is commissioning, producing and reading travel and place writing?

### Literary travel writers and researcher-writers in travel and tourism

The growth in the number of independent travel writers has kept pace with the establishment of city branding as a professional process for destination development since 2000. Destination management departments in city councils, if well-funded, have grown in-house expertise for researching the attributes of their city or have looked outside their public-funded organisations to contract, initially, marketing agencies, to promote inward investment and tourism. Various *Visit* websites promoting cities and urban destinations have emerged (*Visit London, Visit Plymouth, Visit Buxton, Visit Ljubljana, Visit Maribor, Visit Sevnica, Visit Berlin, Visit Bratislava, Visit Tallinn, Visit Copenhagen*). These destination managers also maintain and manage tourism information platforms where travel writing can be found and enjoyed by potential visitors. Since the growth of Masters' programmes in European universities and the establishment of funded research chairs in place branding, the field has seen the appearance of specialist organisations, for example, the International Place Branding Association in the Netherlands, and an increase in the professionalisation of the process of analysis for city brand design. So much so that a city council might well approach a travel writer directly for a project, especially if that writer has created a strong identity and brand around their own writing. The importance of this identity means that the freelancer carefully considers their reputation and strives to build on that in all their work and commercial relationships.

At the same time, easy access to Web 2.0 and a reduction in costs of mobile computing and digital image-making have created opportunities for freelance travel writers to start up their micro-enterprises with low initial investment. Very quickly, staffing agencies have established recruitment

DOI: 10.4324/9781003178781-1

and the re-selling of freelancers' work into the content provision and content marketing spheres of the travel industry. Up-working and gig working can provide freelancers with a portfolio of contract writing, earning at 2022 rates in US and UK £20 per hour, depending, of course, on personal productivity. When reviewing the feedback from freelancers, most suggest on agreeing fees and outputs per project or per job, while another charging method is to maintain a kiloword contract to keep regular clients, who only pay each time a thousand words have been delivered. At 2022 UK rates, this is £450 for a kiloword.

Newspapers and travel magazines, in print and online, are at one end of the market spectrum for freelance travel writers, while travel companies are at the other, requiring text content either for their own websites and promotional materials or as press release copy to send to freelancers and publication editors. Freelancers often join associations or societies in their countries to gain press accreditation and to network for contracts and travel offers, some example associations include The British Guild of Travel Writers, founded in 1960; The International Travel Writers & Photographers Alliance, established in 2003; and The Society of American Travel Writers, re-branded as SATW to hold conventions for meeting colleagues and customers. For longer travel literature and literary travel writing, an independent writer with a book project would be advised to contact a literary agent who specialises in travel. These will advise on fee advances and payment of royalties and on whether the book project will be bought by any publisher or not. An example is the HHB Agency, London W1, but many can be found with an online search. If the book proposal is similar to an existing travel book or author's work, then it is worth trying to find and approach the literary agent who represents that author since they will have the expertise in selling that style or approach.

Travel companies, as mentioned above, and tour operators also need to produce written texts describing and positioning the places they serve. TUI AG, a tour operator, first formed in 1923 as a mining company in Germany, currently, in 2022, employs over 50,000 people to create package holidays, provide cruise experiences and offer flights and hotel accommodation for tourists. Writing and creative careers in the larger tour companies like this are regularly advertised with the following job titles: content manager, content publisher, digital editor, social media marketer, creative and copywriter. Writing positions within the press office or media relations staff of these larger tour operators are also well-established as career pathways. Travel writing skills are, thus, significant in building a career or finding that first job opening after graduation.

## Commissioning place writing in destination development

Agreeing on a contract to write for a DMO (Destination Management Organisation, for example a regional or city council) can be made more straightforward if the steps of the literary travel writing process are formalised with the commissioning stakeholder. The research for this book has developed this process into three steps shown in table form in Chapter 5, as Table 5.7.

The stepped process creates interim outputs which are not only part of the process but also provide auditable deliverables that will show the progress of the project. These interim outputs can be used to communicate progress to the DMO in a commercial project and, in a learning situation, can be shared for formative assessment by a mentor, lecturer or teacher. Guidance for this is given in Chapters 6 and 7.

## Value co-creation in urban space and emerging tourism spaces

In contemporary tourism management, experience co-creation is considered essential to satisfy consumers as experiences represent the foundations of attractions and the hospitality industry. The research by Bounincontri et al. (2017) made on a sample of 385 tourists visiting Naples in Italy showed that interactions among tourists and service providers in the tourism and hospitality industry and active participation of tourists have positive effects on the tourists' level of satisfaction, their expenditures and happiness. With the emergence of new technologies, experience co-creation has been enhanced, and destinations have had the opportunity to facilitate their development and competitiveness (Bounincontri & Micera, 2016).

In smart tourism destinations, particularly, new technologies have been employed to encourage the interaction with the tourists and increase their active participation during the experiences of their stay (Bounincontri & Micera, 2016). Experiences involving tourists' active participation contribute to long-lasting and pleasant memories (Larsen, 2007; Campos et al., 2018) and to personal narratives that are meaningful (Gretzel et al., 2006; Campos et al., 2018). Engaging in activities has the potential to contribute to self-development of tourists in many different ways, from building their identities to enhancing personal competencies, socialising, creating and exploring (Morgan et al., 2009; Kreziak & Frochot, 2011; Mehmetoglu & Engen, 2011; Rihova et al., 2014; Campos et al., 2018). The role of a tourist in co-creating or co-designing of the tourist experiences is, thus, vital (Tan et al., 2013; Campos et al., 2018) and beneficial for many reasons. The researching travel writer, out in the field, is of course, accumulating these same experiences; this section will show later how these can become valuable in the journaling process.

Co-creation has been studied from many different perspectives. What various studies have in common is that they emphasise that increasing interaction with consumers stimulates their contribution to design, creation and consumption of holidays (Räikkönen & Honkanen, 2013; Andrades & Dimanche, 2014; Campos et al., 2018). Co-creation is particularly important in the management and marketing planning of attractions and destinations and has become a significant strategy for achieving competitiveness of destinations in the tourism market (Campos et al., 2018). How to persuade tourists into active participation and how to engage them in co-creation of experiences is the challenge destination managers and product developers try to address.

One of the tools for encouraging active participation, engagement and co-creation of tourism experiences is travel writing, which not only encompasses all the benefits of co-creation but also contributes to the originality and authenticity of the experience which is another very important aspect of the tourism experience. Tourists no longer want to be passive observers of attractions, but they want to be active physically and mentally and become emotionally engaged, and thus strengthen their identity, and competences, and contribute to their personal growth (Bertella, 2014; Campos et al., 2018). Travel writing has the potential to facilitate all that and successfully engage tourists in co-creation of very diverse experiences.

The design of user experiences for smart-phone Apps has become a subject of study in itself in order to improve the user-friendliness of software and to improve engagement with suppliers' websites. Tourism and hospitality service design has begun to draw on the design knowledge of UXD, user-experience design (Roto et al., 2021). This blurring of the two design fields allows researching writers to pick theoretical concepts from UXD to add to their writing design; the written piece here is thought of as a service to its reader-visitors. One example concept that had moved between UX design and service design is touchpoints. These are all the key moments where the tourist makes contact with the service provider. In a hotel, these include the check-in, parking, unlocking the room and, of course, the online booking screen. The hotel strives to make each of these touchpoints a pleasure for the guest. The literary travel writer can imagine their recounted journey across the tourism space of a town to have a series of touchpoints. Their writing will endeavour to provide a memorable experience for each stopping point, where the reader touches the text again. An example of this is drawn from the field preparation for a recounted walk across Exeter city, made during the research for this book. Desk research revealed that the music of the composer, Jackson of Exeter (1730–1803) was still freely available on YouTube. This meant that the researcher could listen to Jackson's sonata before the walk and then take the recording to listen to at a key place in the city, which would then become a touchpoint for the readers during their experience of the text and the route. The writer's work was to find a link between the sonata from around 1773 and a spot in the city. Jackson's music is from the historical period of early Romanticism where sensibility is cultivated and thus lends itself to a literary interpretation by the travel writer.

## Discourse communities in tourism and cultural heritage interpretation

Of course, this book is written from within a discourse community, that of tourism and hospitality design, since its co-authors work, teach and research in that field. The design of the service environment for a tourist attraction, a museum or a hotel takes place before the guest arrives. Buying stock, provisioning, decorating and cleaning are all completed before the visitor experience takes place. Customer-facing specialists and front-of-house staff are the main interface with the tourists during their holiday. One persuasive historical precedent for the introduction of high quality yet transferable service experience design occurs in 1890 when Richard D'Oyly Carte commissioned César Ritz and Auguste Escoffier to launch the Savoy Hotel in London as an initiative to draw the wealthy diners from their homes and clubs to a new tourism space. Providing fine-dining to 1,000 covers stimulated the modernisation of kitchen and hotel management to create a calm, safe and encouraging atmosphere for the new workforce. Escoffier's writing, from the titles he invented for his dishes to the codification of career levels, began the professionalisation of the food and beverage industry. Food preparation need not take place in an environment of bullying, noise and squalor but rather it can be created as a space in which knowledge is handed on and accomplishment is achieved and enjoyed. Escoffier's codification, through writing, served to create this new space. Understanding how people work together to create cultural artefacts and environments is covered in the study of ethnography. Tourism management programmes in higher education prepare undergraduates for ethnographic observation and analysis. The writing-up of these analyses contributes to new knowledge that will make tourism environments safer and more rewarding places to work and at the same time explore how the visitors are co-creating more satisfying experiences.

The travel writer dining in a restaurant, checking into a hotel or queuing for a ride on a tourist attraction can take on the role of the ethnographer. Through interviews, observation and engagement, the travel writer can go some way to understanding what both the tourism staff and the visitors believe they are doing with the artefacts, and this includes designed, named dishes, for example. The travel writer as ethnographer sees both the kitchens and the dining areas as cultural spaces where artefacts are made and used in culturally determined ways. Field observation and dialogue with the users and creators in these spaces will yield important field data for the travel writer. Later analysis of these data from the field can then be incorporated into the travel stories to act as affirmative lessons recounting successful moments of value creation from tourism space. In this way, the travel writer crosses into a relatively new subject discipline, that of ethnography. An older, more well-known subject discipline is history production, writing and consumption.

Educators have explored the requirements for writers to work successfully in subject disciplines, particularly the one mentioned above, history writing (Beaufort, 2004), which is close to the world of literary travel writing. In history publishing, the researchers, readers and other users of the published works belong to what Beaufort calls a discourse community (Beaufort, 2004, *passim*). She endeavours to design a model for educating writers to work in discourse communities, which, she proposes, must take into account 'five knowledge domains: discourse-community knowledge, subject-matter knowledge, rhetorical knowledge, genre knowledge, and writing-process knowledge' (Beaufort, 2004, 137).

In the United Kingdom, two large groups act as the focus for the discourse community in history writing production: the Royal Historical Society, based currently at University College London, and the Historical Association in south-east London, which publishes a more public-facing magazine, *The Historian*. Spend time in any large newsagents in Britain and France and it becomes clear that the discipline of history as a producer of non-fiction narrative has a large public readership and more general acceptance within the public's serious leisure time.

Sven Lindqvist (Lindqvist, 2018) used the travel quest and literary travel writing to create his history of Europe's imperial powers colonising Africa. His publishers classify the resulting book under both history writing and ethnography. However, it is an example of literary travel writing. Lindqvist takes along with him not only his research notes but also the novel of Joseph Conrad's (1899) *Heart of Darkness*, from which Lindqvist draws his title, *Exterminate all the Brutes*. His route can be tracked from his travel story: 7 hours by bus from El Goléa to the oasis town, In Salah, deep in central Algeria. In literary travel writing style, he includes movement, his own emotions and an analysis of deeper historical events that took place in his stopping points or plateaus, for example, the robbing of Alexander Gordon Laing just outside the town of In Salah. In towns, he takes time to describe the built heritage, for example, the Sudanese style with its white pillars that enliven the reddish-brown clay. Then, he offers interpretation of this heritage for his readers.

In Britain, large employers in heritage interpretation include the National Trust, with a staff of about 14,000 plus 50,000 volunteers, and English Heritage with 2,700 staff. These form the basis of a community that understands why a society preserves, communicates and values built heritage. Along with those who work within the heritage world, the discourse community also includes researchers and academics who specialise in the history of urban space, architecture, art and landscape. Finally, the interpretation process is so well developed in the twenty-first century that large numbers of the general population visit heritage properties as a leisure activity and read travel and heritage texts linked to these sites, for example, the National Trust regularly reported 5 million annual visitors up until the COVID pandemic began in early 2020.

Literary travel writers then, who create texts for this discourse community, need not only the specialist knowledge to communicate their findings with their academic peer group but also writing skills that will reach a wide, yet interested public audience. An early example of this type of publicly accessible, but learned, writing was the work of the pioneer area studies academic, W G Hoskins (1908–1992), who unearthed narratives in the landscape of the English counties that he studied, for example his tome called simply *Devon* (Hoskins, 2003), which has been reprinted since its first publication in 1955. The lecturer in German literature, W G Sebald (1944–2001), turned to literary travel writing with his second book, *Vertigo* (1990), followed by three more travel books in the mid to late 1990s, which appeared first in German and then in English translation. While Sebald's *A Place in the Country*, original German 1998, and in English 2013, is closer to literary essay in style, it still shows a concern with the places inhabited by writers or touched by literary sensibilities and still connects with a more general readership than an academic article would do. In *Rings of Saturn*, Sebald uses a first-person narrator and very consciously mobilises the book by recounting a walking tour in East Anglia. Thus, book publishers have become more aware of a genre of literary travel writing that may be consumed as much for the tourist destinations it depicts as for the following the author has gathered who find pleasure in the reading of the text.

In encouraging literary travel writers to interrogate and open up their own self-identities as they develop their knowledge the field turns to the diary, and indeed, the gerund form, journaling derived from keeping a daily log. While a discourse community does exist for this, it currently focusses on French diary-keeping, the APA, the Association for Autobiography and Autobiographic Heritage; the subject of keeping day-to-day thoughts has now gained attention from academic research. Sam Ferguson (2018) introduces diary-keeping and the critical responses to this form of writing in a recent study that provides a very useful opportunity for literary travel writers to consider audience when journaling for a commissioned project. A new self-awareness and concern with witnessing everyday life through writing are attributes of modernist literature from Joseph Conrad (1902) *Heart of Darkness*, Virginia Woolf (1925) *Mrs Dalloway*, Italo Svevo (1923) *Zeno's Conscience*, Katherine Mansfield (1918) *Prelude* and Hope Mirrlees (1920) *Paris: A Poem*, through to André Gide (1925) *The Counterfeiters* with its companion behind-the-scenes *Journal of the Counterfeiters* (1927). Ferguson's reading of Philippe Lejeune's extended work on the diary raises questions of truthfulness, literariness and perceived audience for the private diarist. These considerations help to problematise the stages of journaling for the travel writer.

As the days take shape in the notes taken during a journey, this structuring, although convincingly hidden behind the passive voice here, is shaped by the travel writer. The style, the tone, and the personal attitudes all demand attention. The naivety of simply bearing witness to the truth of the events of that day is challenged immediately by the writer's selection of what to record, what to report upon and in what light to present it. Gide, in *Paludes* (1895), reminds readers that he is not just journaling as he synthesises his writing for publication, but is aiming to add style. So much so, that he invites readers to collect quotable phrases from the novel and to journal them themselves in a space he asks his publishers to leave at the back of the book. Modernist literary authors, like Gide, offer ideas for experimentation for contemporary travel writers and place branders. It is a full awareness of the collection and fashioning of the text and also the understanding of its delivery and its readership, that these experimental novelists draw attention to.

## The community of travel writers and their market

The discourse practice community as a working body of writers and creatives in industry has had opportunities to meet peers and buyers since at least 2009, when Travel Blogging Exchange, TBEX, first organised a conference in Chicago. TBEX aims to attract 500 travel content creators each year by offering training in writing skills and monetising. It also draws commissioning companies from tourism and destination management, from which it can offer sponsored familiarisation and marketing, FAM trips to creatives. TBEX runs its conference annually in the United States, Europe and Asia and has re-started post-pandemic.

TravelCon is an events organisation in the United States that has developed meet-the-buyer conferences for bloggers in the United States, too. Both conference organisers invite speakers with travel writing portfolios, for example, Faith Adiele. Adiele's writing practices demonstrate to the discourse community the range of different media in which their writing creativity can earn them a living, for example, podcasting, pay-per-click advertising on a regular blog, YouTube video-blogging, book-length travel literature and winning magazine commissions.

Familiarisation trips, FAM trips, extend the discourse community not just because the attendees on the FAM trip will be a group of bloggers and writers, but because the trips are organised by career public relations professionals in larger organisations who are also writing about tourism place. The PR staff write and maintain digital content on place, themselves. An example of this is RTO 9, the regional tourism organisation funded by the Ministry of Heritage, Sport, Tourism and Culture in Canada. RTO 9's region is South Eastern Ontario and their web pages explain their investment in PR staff and in their media familiarisation work. These regional destination organisations might also be in a non-Anglophone country but still aim to attract English-speaking inbound tourists and thus are always trying to attract writers from this target language group. One example, which was part of the research for this book, is the French-speaking region of Wallonia in southern Belgium. Writers who are innovating with their writing practices need to read a constant supply of new writing to stimulate their growth and need to see how others are learning to perform their writers' practices (Weber et al., 2007).

Literary travel writers cover more than just the requirements of the itinerary in their stories and this means they will find value in accessing the specialist sub-groups in the writing world. These specialists have often organised themselves into associations and guilds with websites for membership and include the Circle of Wine Writers, the Guild of Food Writers and the Heritage Management Organisation with its headquarters in Chicago. Architectural writers also have at least two membership and training organisations in the anglophone world. While the Critics' Circle in London has members in the following sections: books, music, drama, film, dance and visual arts, which provides travel writers with a useful list of which cultural expressions attract appreciative writing. The Music Critics Association of North America has been in existence since 1956 to run workshops and conferences on music appreciation writing, which will be useful to the travel writer when linking their urban experiences to music. Travel writers also need an awareness and sensitivity to ethnobotany for which garden writing associations provide a peer group and discourse community, for example, the recently re-branded GardenComm in California. Some of these societies will provide press-type access to companies and their press offices for more direct research and fact-checking that the general public cannot access.

## University education in literary travel writing, creative writing and place branding

In tourism management education, one of the key disciplines at graduate level teaching is social anthropology and in particular ethnography within that sphere of study. This discipline, too, has its discourse community represented by interest groups and associations. These are mainly academic, rather than market-facing as the societies covered above. In Britain, the Association of Social Anthropologists of the United Kingdom, the ASA, has a membership facility and organises open conferences and what it calls, studios, to continue the dialogue throughout the year. In 2022, for example, the ASA ran studios on decolonising the institutions of the university in the west. The Royal Anthropological Institute of Great Britain and Ireland (RAI) also offers membership to practitioners and academics. Travel writing from the nineteenth century shares its history with the creation of anthropology as a science studied at universities in Europe and the United States. The European Association of Social Anthropologists, the EASA, continues to publish work in this field of research in its journal *Social Anthropology Anthropologie Sociale*, while the European Anthropological Association, the EAA, continues this work from its base in Hungary.

The American anthropologist, Paul Rabinow, re-established the link between narrative non-fiction and anthropological field reports with his *Reflections on Fieldwork in Morocco* (1977) in which he uses the literary devices of character, story, emotion in a first-person hero along with intrigue and plot to retell his scientific study of Sidi Lahcen, a Berber town and the city of Sefrou, in the Middle Atlas Mountains. While his literary travel account worried anthropologists who had striven to gain acceptance of their discipline as a positivist science, travel writers were reminded that their work was inquiry and valid knowledge creation. Something stronger than a mere trace of an enquiry is visible at the heart of Rabinow's book; analysis is visible, the link to the socio-political fallout of French colonialism is made explicit, and his narrative re-telling is the synthesis required of scientific study. The study does make cultural conclusions, too, around one of anthropology's key questions: is this a group of people in a state of successful social reproduction or is emergence of a new social order being witnessed? The detection of a new type of recolonisation is an enquiry that the contemporary literary travel writer can explore through their writing as inquiry.

The risk that Rabinow and his publishers took in 1977 was vindicated over the next three decades culminating in the publication of a celebratory 30th anniversary edition. In hermeneutic research, the interpretation of the journaled experiences comes later. The understanding is not there at the time of the experience. Thus, a cultural act finds its meaning in what comes after it. This was true of Rabinow's travel writing and place-making, and it serves as encouragement to contemporary travel writers to go out into the field and record their experiences, even if the problem is not entirely clear.

## Conclusion

This introduction has expanded on the disciplines and cultural practices that travel writers can bring into their work and at the same time shown where dispersed communities of support are available to help them to engage with the specialisms but also to find established niche markets for their work. It has also shown that a larger upsurge of writing opportunities exist in city branding and destination management and that this emergent field is just beginning to formalise itself. This book opens this exciting field further by proposing methods of working with larger organisations and developing the understanding that writers can be professionally commissioned by building trust with these bodies. Finally, it is suggested that literary travel writers can sensitively change their identities and grow their confidence as they work and create their portfolios. This innovative processual method is examined in detail in the chapters that follow.

# 1

# INTRODUCING LITERARY TRAVEL WRITING AND PLACE-MAKING

## Urban space as a place of happening

Academic research publications on literary tourism show that literary writing influences and informs perceptions of visitors and tourists and their wish to visit certain places. In literary tourism, the concept of 'text-to-tourism' is often inverted and changed to 'tourism-to-text' (Potočnik Topler, 2016), especially concerning urban places. Norman Mailer is among authors, who extensively described urban places and used them as 'places of happening' in his novels. Some of the cities that Mailer described in detail or just briefly mentioned are Washington, Chicago, Miami, New York, Los Angeles and San Francisco. Mailer's *The Armies of the Night* (1968), which the *New York Times* critic, Kazin labelled as 'diary-essay-tract-sermon' (Potočnik Topler, 2016, 50), is set in Washington, where the anti-war demonstrations took place in 1967 from 21st to 23rd October. Mailer's descriptions of events are very detailed and vivid, and what is especially characteristic of this Mailer novel is the blurred line between fiction and journalism. When it comes to style, *The Armies of the Night*, which is comprised of Book One: The Steps of the Pentagon, subtitled History as a Novel, and Book Two: The Battle of the Pentagon, subtitled The Novel as History, is linked to literary journalism, New journalism and participatory journalism. According to Mailer (1968, 284), Book One is 'a personal history which while written as a novel was to the best of the author's memory scrupulous to facts, and therefore a document'; and Book Two is in its form similar to a news report or reportage. As journalists do, Mailer employed a variety of different voices in his novel and also different discourses. His novel *The Armies of the Night* is therefore an example of heteroglossia, the employment of various discourses in a particular work.

DOI: 10.4324/9781003178781-2

In *Miami and the Siege of Chicago* (1969), Mailer is a reporter, documenting the political arena in the United States at the end of the 1960s. In the novel, he successfully employs urban places and speaks of Chicago and other American cities with pride and enthusiasm:

> Miami is the great American city. New York is one of the capitals of the world and Los Angeles is a constellation of plastic, San Francisco is a lady, Boston has become Urban Renewal, Philadelphia and Baltimore and Washington blink like dull diamonds in the smog of Eastern Megalopolis, and New Orleans is unremarkable past the French Quarter. Detroit is a one-trade town, Pittsburgh has lost its golden triangle. St. Louis has become the golden arch of the corporation, and nights in Kansas City close early. The oil depletion allowance makes Houston and Dallas naught but checkerboards for this sort of game. But Chicago is a great American city. Perhaps it is the last of the great American cities.
>
> *(Mailer, 1968, 83)*

New stylistic voices appeared here in Mailer's approach to story-telling encouraged by the experiments of, for example, Truman Capote's *In Cold Blood*, where the narrator takes neither first person nor appears as a third person. Subsequent writing about urban space has been profoundly affected by these experiments from the late 1960s in American English. Mailer's novel *Why Are We in Vietnam* associates these narrative experiments with travel writing as the author recounts a hunting trip in Alaska that was made by a group from Texas. A specific set of detectable practices within the written text communicate the impression that Mailer is writing from a real place and that he is there in that place sending that message to the readers. Readers far away in Europe are convinced by the text that this place called America exists and can be visited. It is a use of language so deeply embedded in this western discourse that readers will not even stop to question the existence of the places that Mailer creates with his writing practices. Indeed, even the critical reader will not question whether, for example, Los Angeles exists but might engage at a direct level with Mailer's description of the city and attribute his opinion of it to his identity. At the fundamental level, western users of English can detect the messages in this type of place description and the form in which Mailer's publishers present it, to believe that these places can be visited.

### Writing practices in tourism, travel, place branding and place-making

When the term deltiology, appeared, from the Greek *deltion*, a small writing tablet, in 1945, postcards were already established as a writing practice for visitors at trade fairs. Postcards communicate holidaymakers' identities through their own written messages to friends and family back at home along with the destination's brand on the card's picture face. Telling others about places through writing has a longer history than the nineteenth-century emergence of postal technologies and infrastructure but these new technologies allowed the practice to flourish among a broad range of social classes. In this book, eight contemporary writing practices are examined in detail as they create, recount and refract the places that people explore and share. By examining these travel and place-writing practices and their interaction with socio-economic development, it is hoped that a better understanding of how identities are formed and how places are created will be reached. Beyond that though, the study is an invitation to engage in place-writing practices, to achieve well-being and to see how they might open up fields of inquiry and contribute more complex and problematised systems of knowing in an increasingly literate society.

## Promotional tourism copy

Beyond the postcard message of 'I am enjoying my holidays' or 'Wish you were here', the first major category of travel writing in the tourism and hospitality industry is the travel brochure, which has migrated onto the web both as screen content, interstitial advertising and as text in pdf brochures. The stakeholders of this discourse practice are clear both to the writers and to the readers, who are its consumers. The travel company wishes to explain and advertise its products and copywriters, or content authors and creatives use a coded language (grammatical structures, words and writing techniques are carefully chosen) to promote and exhort the company's offer, its values and desired narratives. Professional communication is essential since visitors and consumers with quality information are able to pick suitable and rewarding experiences for themselves and display mindful behaviour towards the service providers and the local community.

Positive representations of destinations are achieved by employing various linguistic devices and techniques, for example Dann's (1996, 101–134) tourism categories, to which he refers as three Rs (Romanticism, Regression, Rebirth), three Hs (Happiness, Hedonism, Helio-centrism), three Fs (Fun, Fantasy, Fairy Tales) and three Ss (Sea, Sex, Socialization). Dann (1996, 68) also discusses four sociolinguistic models for promotional texts in tourism, which are divided into the following four categories: (a) the language of authentication (promotes the experience of the traveller as authentic, genuine, pure, opposing it to the banality of everyday life); (b) the language of differentiation (highlights the contrast between holiday and normal life); (c) the language of recreation (emphasises the recreational and hedonistic side of tourism); and (d) the language of appropriation (tries to adopt an attitude of control and domination of what is unknown).

## Guide book

The guide book for tourists has been established since the 1820s, with the introduction of a series of traveller's guides now referred to as Baedekers, named after the original publisher and entrepreneur.

Professionals, preparing the contents of guides have become increasingly aware of the role of communication and discourse in tourism, especially in the representation of destination towns. Tourists are 'actively seeking simulated, fun-filled experiences to meet their expectations' (Pabel & Pearce, 2018, 85), and this is carefully considered by guide-book writers. Destination descriptions in these books are very close in style to promotional texts, so they are communicative and easy to understand. Their main functions are to encourage, to inform and to instruct, that is, descriptive and directive types of text. They are directive in that they can tell the reader what to do next, occasionally the writer will use the imperative mood of the verb to do that, for example: try the coffee at one of the street cafés in the main square. In destination descriptions, where the use of present tenses prevails, the present simple tense is the most common, usually simple sentences are used interchangeably with complex sentences to give pace and variety to the text. This is achieved with coordinate and subordinate clauses and in the latter relative clauses, conditionals and concessive clauses (Potočnik Topler, 2022). In the extract below, from a Rough Guide to Trieste, directive phrases address the readers, telling them where to go to experience traditional tourist activities. The present tense creates the impression that the streets here will always offer those same activities, so much so that the author has almost added a disclaimer, by telling the readers that there will be changes in the bars:

> The best area to head for is pedestrian-only Viale XX Settembre, known as the Acquedotto ("aqueduct"), where citizens stroll in the evening. Via C. Battisti, east of Sant'Antonio, is good for food shops. For late-night drinking, Via Madonna del Mare, below the castle, has a number of bars whose names, managements and popularity come and go each year.
>
> *(Dunford, 2013, no page numbers)*

### User-generated content in reviews on social media (UGC)

Described in more general marketing terms as unsolicited user reports, and solicited user reports, reviews on TripAdvisor and on the Google Maps' local guides' scheme form a writing practice that is very visible at the point-of-sale for holiday destination choices and during stays in cities when additional activities and attractions are sought and booked.

### Travelogue

In one of the groups at the University of Maribor, Master's students were offered a choice of tasks using travelogues, with the goal of improving communication skills in the English language and to practice tourism discourse techniques. Travelogue writing is a prerequisite for offering high quality services, well-communicated products and professionally branded destinations. It was emphasised during the assignment that not only proper language skills in the mother tongue and in foreign languages, especially in English, are important in tourism and hospitality careers but also awareness of cultural place specifics, which include heritage, literature, history, language specifics, terroir, food and climate. All these skills are essential for communicating with consumers on the one hand, and to present attractions and develop the destination image on the other. In the assignment, students tried to combine their knowledge and creativity and produced scripts in duration from 4 to 12 minutes while enjoying a literary walk at a certain place, destination or attraction of their choice.

When discussing travelogues, some theoretical and methodological issues arise. The first one is probably the definition of the travelogue. Literature overview offers various definitions. One of the reasons for that is that there are many types of travelogues, which are often related to travel motives, travelling for personal reasons or travelling for business reasons. Research has helped to understand diverse motivations, emotions and reflections that accompanied traveller's itineraries (Lucchesi, 1996). The internet site entitled Travel Writing World explains a travelogue as 'a truthful account of an individual's experiences traveling (*sic*), usually told in the past tense and in the first person' (travelwritingworld.com). The coinage comes from the words originating in Greek – *monos* (alone) and *logos* (word, speech) (travelwritingworld.com). Even as promotional texts travelogues have the potential on the one hand to inform, persuade, remind and entertain the readers, and on the other, to be innovative and creative.

*Journalistic travel writing and travel blogging*

Often framed as the author's review of a visit to a city, resort or attraction. National newspapers, both in print and online, carry regular sections usually headed as travel to create a space to sell advertising from hotel, travel and tourism companies. The advertising revenue affords the publishers funds to reimburse the travel journalist, either in a full-time staff role or as a freelance contributor. Additional funding-in-kind for undertaking the journey and hotel accommodation at the destination is often provided by the DMO, the hotel and the airline or ferry company. Although journalism has a long tradition of protecting its sources and of remaining independent of financial inducements, readers generally accept the funding structure for these travel reviews. With the move online, these travel reviews now have a longer lifespan that the printed issues of, for example, Sunday newspapers and colour supplements.

With the rise of travel blogging since the advent of Web 2.0, and in particular since the emergence of low-cost content management systems, including WordPress, freelance writers are now able to attract their own advertising revenue. Schemes, for example, AdSense via the digital advertising company, Google, allow more technically adept freelance travel journalists to monetise their blogs. A regular flow of digital travel content, which readers value and return to, will draw advertisers' interstitial links onto the blogger's pages; if clicked these followed urls earn micro-payments into the freelancer's Google account. When discussing bloggers and travel writers, sometimes good travel writers can be bloggers, but it very often happens that bloggers are writers with less experience, smaller audiences and less income from writing. There is no clear-cut distinction, though and no rules about who is a better writer – travel writers for established newspapers and magazines or bloggers, but major newspapers and magazines usually do not engage poor writers while basically everyone can start a blog. Very often travel writers for established media are journalists with journalistic courses and bachelor degrees.

### Non-fiction literary travel writing and place-making practices including autoethnography

This approach to writing, which is developed fully in this book, is to propose and use writing as a method of inquiry. In particular, it is applied to urban space and heritage tourism sites here as a way of presenting a model for future use. The journey, hotel accommodation and entry fees to attractions and museums for the fieldwork research are increasingly accepted as legitimate support by a research council or funding body, and the author declares this support with any publications in the academic literature.

This category also contains academic ethnography written up and published, and travel literature researched and written by authors practising as travel writers professionally; these two practices form an intersection within this category. The writing methods, and even the methodologies of fieldwork and semi-structured interview, are often used by any of the three types of writers in this category.

### Accidental travel literature, often fiction

The category of accidental travel literature comprises travelogues written unintentionally during the trips or after the trip has finished. Such texts can be published online, in daily, weekly or monthly newspapers and magazines, and in books, as fiction or non-fiction novels. Among authors of this genre is Slovenian author Branko Gradišnik, who wrote several travelogues titled *Strogo zaupno na Irskem* (*Strictly confidential in Ireland*), *Strogo zaupno po Siciliji* (*Strictly confidential in Sicily*), *Strogo zaupno o Sloveniji* (*Strictly confidential about Slovenia*) and *Strogo zaupno na Portugalskem* (*Strictly confidential in Portugal*). In his texts, Gradišnik as a knowledgeable observer describes customs, history and other peculiarities of the visited destinations by employing humour. Among world famous authors in this category are Bill Bryson (*Neither Here nor There*), Paul Theroux (*The Great Railway Bazaar: By Train Through Asia*) and Tony Hawkes (*Round Ireland with a Fridge*).

This is an excerpt from the travelogue *Strogo zaupno na Irskem* (*Strictly confidential in Ireland*), translated into English:

> Ireland is a land of unmemorable and easily interchangeable places for our ears, and a traveler who, for example, travels from Ballymen via Ballyclare and Banbridge to Ballybay and on to Bailieborough near Ballyjamesduff may by the way find themselves in the backwater of Ballynahinch or Ballygawley. As a rule, traffic junctions are arranged in islands, from which there are five or six roads, and in front of each island there is a signpost with five or six unspeakable names. There are some legends circulating among tourists about heroes of their kind, who were lucky enough to read all the names, but the locals will be able to tell you, in a moment of pub confidentiality, many horrible stories about drivers who tried to read the signposts, and in doing so, they stopped looking out for other road users, and consequently made a fatal car accident, and now the punishment ghosts of the deceased in these crashes visit the fateful traffic island again and again and circle it with phantom vans doomed to eternal torment of reading the same Irish place names behind the circle.
>
> (*Gradišnik, 2001, 49–50*)

## DMO strategy documents

Through social media the visibility of destinations, attractions and local businesses is increased. An example of putting a new destination on the tourist map is the Slovenian town of Sevnica. The recent story of tourism development in Sevnica, a municipality with around 5,000 inhabitants, began in 2016, when the rural town on the banks of the Sava River started appearing in the world media because Donald Trump's wife, Melania originates from Sevnica. Initially, the community of Sevnica felt pressure from this sudden intrusion. Journalists were wandering around, asking questions and invading the privacy of some locals. Eventually, the community tried to be supportive of the media, and it saw reporting by the large press agencies (Reuters, ABC, BBC, CNN, NBC, *Washington Post* and *Frankfurter Allgemeine Zeitung*) about Sevnica as a great opportunity to promote their tourism offer. The tourism stakeholders, in particular, saw the need to implement not only traditional but also new media, which turned out to be vital for presenting and developing the offer, new products and future tourism of Sevnica in terms of sustainability. It is the new media and new technologies that enable information transfer in the first place, and thus, tourism branding, tourism marketing and customers' feedback. By engaging new media diverse promotional techniques may be employed, to inform the users about sustainable practices and the preservation of the natural and cultural heritage of the destination. A step further is the creation of platforms for exposing tourists to the heritage, traditions, creativity, language and art. Since 2018, Sevnica has been a member of the Green Scheme of Slovenian Tourism which is a certification programme developed at the national level that brings together all efforts directed towards the sustainable development of tourism in Slovenia with the main goal to bring sustainable models to both tourism service providers and destinations in Slovenia.

These national and regional efforts by government departments result in the writing and storing of a type of travel literature that creates writing work for professionals but also acts as a source archive for future travel writers and journalists who want to check facts and find specific attractions. Almost all destinations now maintain these strategy documents, often for public access via the web, for example, Slovenia's Ministry of Economic Development keep updated tourism literature at this reference (Slovenia, 2017). Smaller communities can access these and use national initiatives to guide their own tourism development at a local level by finding synergies.

### Non-fiction literary travel writing and place-making practices

In trying to propose a definition for non-fiction literary travel writing, it is identifiable by its readers through four key features, movement is in evidence, the past tenses are employed signalling a completed narrative and a first-person narrator guarantees the ethical dimension of eye-witness testimony. The fourth feature is the immanence of the story; all that the readers have is there, and this total testimony is evidence that the narrator lived to recount the tale.

### The six past tenses in English narrative writing

Past Simple: I walked, I was. We were. I went.

Past Continuous: I was walking (sometimes called the past progressive in US English).

Past Perfect: I had walked

Past Perfect Continuous: I had been walking

The habitual form of the past, prefixed by 'used to' or the auxiliary 'would' which reads like a French imperfect tense e.g. 'I used to walk along this street when I was a student'. Or 'I would walk through the park every day'. Note, this latter is not conditional upon anything even though 'would' is used.

The nostalgic Present Perfect, for example, 'I have lived here since I was a teenager'.

The Past Emphatic may also be used, e.g. 'I did walk down that same street'.

The aim of writing using past tenses is to insert the presence of the travel writer into the space and the story, to create a narrative and to authenticate the journey that was undertaken.

This example from Sebald's *Vertigo* in English translation (Sebald, 1999) exemplifies these defining features:

> The whole of the darkly gleaming lake lay silently about me [...] I lit the lamp in the stern of my boat and set myself rowing again, half towards the western shore and half against the cooling northerly breeze that passes over the lake every night.
>
> *(Sebald, 1999, 92)*

Notice the verbs in the past simple tense, lit, set, even the lake lay silently. Two agents are in motion, too, the narrator rowing and the breeze passing over Lake Garda. The first-person narrator is using the personal pronoun 'I' and even takes possession of the borrowed boat with the determiner, 'my'.

## The growth of autoethnography

Separating the researcher's identity into a data-gathering producer of a field journal suitable for later analysis by the same researcher is incorporated into the methodology now known as autoethnography in the study of destinations and leisure practices (Coghlan & Filo, 2013). Gale (2018), developing writing as inquiry, puts forward that autoethnography means 'that there can never be a view from nowhere' (Gale, 2018, 87). Gale takes the spatialising concept further, which is a useful metaphor for travel writers to think with: 'We are positioned, we have a stance [...] we therefore look at the world through particular lenses' (Gale, 2018, 87). The stance or position of the narrator and the literary text is explored by Althusser (1971) in his notion that the literary text hails its reader, the term is interpellation. The narrator is the centre of deictic reference, when the narrator says 'from below me, the sounds of traffic on the quayside came up to the hotel window', readers are centred there with that first-person entity. Indeed, the deictic prepositions, up, below, pre-position that character as the narrator rather than as a third person. The point from where the enunciation emanates, where the narrator stands when speaking has the Greek term, the *origo*, although no term has been proposed by existing academic literature to describe the position of the recipient, in classical Greek this would be σκοπός – *skopós*, the stance of the interpolated one. The enunciation travels from a point of origin to a destination, a process imitated in travel by the tourist.

This type of research-based travel writing often includes the qualitative method of autoethnography. Autoethnography often stems from the researcher's environment (i.e. researching topics and areas that are personally related to the researcher) and is inclined to subjectivity. Ellis and Bochner (2000) discuss autoethnographic writing and evocative autoethnography and argue that evocative autoethnography requires considerable communication skills. Anderson (2006, 375) uses the term analytic autoethnography in reference to

> ethnographic work in which the researcher is (1) a full member in the research group or setting, (2) visible as such a member in the researcher's published texts, and (3) committed to an analytic research agenda focused on improving theoretical understandings of broader social phenomena.

Anderson (2006, 378) argues that analytic autoethnography has the following five key features: '(1) complete member researcher (CMR) status, (2) analytic reflexivity, (3) narrative visibility of the researcher's self, (4) dialogue with informants beyond the self and (5) commitment to theoretical analysis'.

However, the methodology is being used in the research field of tourism and hospitality management. A recent example from Lee and Ruck (2022) takes the autoethnography further into a narrative writing of the findings and even synthesises composite characters from the industry to communicate these findings:

> Scene 2 [...] At 7:15am, the city gradually awakens from its slumber. Exhaust fumes from rumbling car engines fumigate the city streets. Storefront employees raise and rattle the chains of their shutter gates. The rapid-transit stations burgeon with human activity, producing a wave of pedestrians outside the doors of the coffee shop. Before long, customers begin entering [...].
>
> *(Lee & Ruck, 2022, 9)*

At the end of each of the scenes, reproduced in the research article, the authors do return to the themes, to discuss them further, using, for example, the theory of habitual powers and how they foster artistic-aesthetic experiences (Lee & Ruck, 2022).

## Twill as research synthesis

How then can travel writing structure narrative to simulate not just space but also construct movement from one plateau of place to another plateau? Modiano (1989) offers a method in his novel, *Vestiaire de l'enfance*, using literary language. He, or his narrator, alerts readers that he is considering literary value (Modiano, 1989, 10) in novels versus the radio serial, using the imagery of *tissu* in French; a woven twill or fabric would translate this into English, no English translation of this novel exists at the time of writing. In a twill weaving structure, the threads are continuous, viewed in medium close-up these appear as individual points, consider for example dogtooth check. However, from a distance, a pattern appears called the wale. Each tooth of the check rhymes with the one at the end of the next line. In twill, these points simply reveal the floats, where the darker yarn floats above the pale yarn beneath. In the novel, Modiano then produces a twill in his writing over the next two scenes or plateaus from a travel-writing stance. First, in the radio station, a character is observed drinking a local mineral water, which the narrator explains has a peculiar taste (Modiano, 1989, 11). The action moves to a new scene, in a new location, the Rosal Café, and the author uses the local mineral water again. It is a rhyme with the practices of the first character. It tells the reader that this next character is already crossing the boundary between tourist and ex-patriate. These floats, rhyming points using everyday practices can be used to produce literariness in travel writing by providing contiguity between disparate scenes. The contiguity communicates movement or travel.

Twill contiguity can also be used as a research approach for this type of travel writing that moves from autoethnography to ethical connection with locals. The next section examines a contiguity section from Sebald's (2002) *Vertigo* travel book to see connection with locals by the travel writer. This can form the basis for later ethical research, as the following section explains. First, though, consider the routes taken by travel writers and the stopping places which are created by their writing; these are referred to as plateaus. A plateau can be a café where the researcher-writer is collecting data in note form, or even beginning to write up these notes in a form close to the text to be published. A series of scenes from Sebald's (2002) *Vertigo*, exemplifies this type of plateau. Max Sebald, the narrator, is staying at the Hotel Sole in Limone sul Garda, north of Verona in Italy. From the suggested dates in the text, it is around 1st or 2nd August 1987 (Sebald, 2002, 91–95), and see date on passport (Sebald, 2002, 114).

Sebald's structuring of his plateaus provides a model for exploring contiguity. Once understood in these terms, it then offers an opportunity for a research testing to be implemented into literary travel writing for practitioners interested in ethics in their writing. Three scenes, or as this work calls them writing plateaus, unfold in sequence in Sebald's stay and subsequent departure from the Hotel Sole. The first scene is a true writing plateau because Sebald says he is writing the travel literature that the readers are reading 'I sat at a table near the open terrace door, my papers and notes spread out around me' (Sebald, 2002, 94). Notice how he is stationary and has access to a flat surface, a plateau on which to write. During these two pages, he describes a conversation between himself and Luciana Michelotti, the hotel owner, who is also working, serving snacks and drinks. Luciana asks if the town of Limone will feature in Sebald's story (Sebald, 2002, 95).

The second scene, in which Max and Luciana are parting, is next to the bus stop for Desenzano; Max has found a plateau again where he 'ordered an espresso, and soon became [...] deeply absorbed in recasting [his] notes' (Sebald, 2002, 103). Luciana, by being incorporated again, has become a character in which Max expresses an emotional connection. This propagation of a character over two or three plateaus echoes the theory of contiguity outlined above.

The third plateau is on board the train, much later, to Milan. This third scene is the line of fleeing from that place. It is difficult to discern if Max is still thinking in emotional terms about Luciana, although he is remarking on other passengers. He says he is 'visible again to [his] mind's eye' (Sebald, 2002, 103), but it is unclear when he is taking notes for this on-board plateau. This sequence of scenes offers researcher-writers an opportunity to return to their respondents in a writing inquiry project to open to their respondents the possibility of expressing their relationship with place. A researcher-writer emulating these scenes from Sebald could stay in contact with the figure represented by Luciana and write to ask her if anything was missing from the plateaus in which she figured. This approach is an attempt to better understand what others' practices mean to them in these places. It may have to be phrased in more direct questions to elicit any response, e.g. why do you …? What did you do that for? The responses, once received, could then be incorporated into the fleeing scene as a propagation towards a literary text.

Even without this correspondence with a respondent after the fieldwork, twill offers a method of interrogating a topic. The researcher-writer can return to the topic of investigation by deliberately adding points, like the floats of a twill pattern, into their recounting of the scenes from their journey. At each point or float, the travel writer turns over the question or proposes a new understanding of the question from the topic. Seen from a distance, these floats become the wale and the reader's experience is a pattern in which the researcher considers a topic in a range of smaller inquiries; please see example below extracted from the Belgian place-branding blog. The topic being investigated and brought to the fore in repeated considerations is the short-cut taken when passengers are travelling for work rather than for pleasure and tourism:

Lille to Liège – Desire Paths 14.12.20. Wallonia regional branding weblog

Max Sebald once said that the clocks in Lille and Liège kept the same time but were different from the other cities across Belgium in the nineteenth century. It was only the connection by train between French and Flemish speaking cities that brought the times into synchronisation. I was a visiting lecturer at the University of Lille 3 in the 1990s. I used to take the train from Newcastle station when I travelled to teach there and then change onto the old Eurostar service from London Waterloo. The Eurostar train only started daily services in May 1995 so it was still an exciting novelty to see Lille shown on the departure boards far in the northeast of England. For a short time in the 1990s, the incoming train from Newcastle would stop out in a vast yard of rail lines in London, the engine driver would disembark and walk slowly the full length of the train then start again in reverse to complete the last few miles into Waterloo International.

My stay in Lille during each teaching period was always intensely busy so my exploration of the city was limited. I grew to know the Allée de Liège, with its glass shopping mall, Euralille on its southeastern side. If I arrived late in the afternoon on the Eurostar, then the mall provided a quicker walking route than the allée and offered the additional benefit of a small coffee bar just before the exit. On my later stays, I started using the Hotel Lille Europe, which was part of this gigantic glass mall. Travelling for work always creates economies of movement.

At breakfast, the room-height windows of the first floor offered a panorama of the city. The tinted glass cast a bluish grey cloud over the commuters as they hurried to work. North-west across the double width road, I could see the Henri Matisse Park being laid out; it was still a contest between new paving and those desire paths trampled by workers late for the office. I did not realise back then that Wallonia lay due east of the Allée de Liège and that a desire path began only 200 metres from where I enjoyed breakfast.

## Ricœur's *mythos* and accretion of culture

William Dowling builds a new figuration of how literary works contribute to the culture of a place by drawing on Paul Ricœur's writing on narrative (Dowling, 2011). Dowling renews Ricœur's *mythos* concept in this way:

> What then comes to light is an alternative notion of *mythos* as what Ricœur calls an arc of operations, a complex movement that originates in culture understood as a symbolic order, that then passes into fixed or frozen form in a work like the *Iliad* or *Middlemarch*, and that is then finally reintroduced into the cultural sphere in the consciousness of listeners or readers whose way of being in the world has been altered by their reading.
>
> *(Dowling, 2011, 2–3)*

A new metaphor is introduced, that of the accreted archaeological layer, this notion of the arc of operations sheds light on literary destinations which are mediated over several eras. A literary author may be the first to write about a true event that takes place at a specific location using the event for dramatic effect in a story, or use local climate knowledge to add realism to a scene in the work of fiction. The example of the changing wind directions over Lake Garda is used by Goethe and Stendhal in their writing, but then Kafka and later still Sebald excavate these literary strata to incorporate the winds and their effects into their travel stories. What is important from Ricœur, via Dowling's new reading (Dowling, 2011), is that readers have had their way of being in the world affected by these narrative accretions. These readers in turn may become visitors or travel writers and continue the laying down of new strata in the archaeology of the symbolic order of the place. Both the literary travel writers and the stakeholders of a place that has literary accretions face the same question: will visitors know this author? Sebald makes his readers work a little harder to find who the author is on Lake Garda by using only Stendhal's real name rather than his *nom de plume*. Sebald's approach maintains the mystery so that literary tourists have something to solve for themselves at the destination. At the other end of the spectrum for literary enthusiasts is the former home of Agatha Christie in Devon where the on-site bookshop presents all her detective novels for visitors to browse and buy.

## Letting others speak in literary travel writing

Rabinow's (2007) travel book drawn from his anthropological fieldwork in Morocco in the late 1960s was mentioned in the introductory chapter as an example of the social scientist crossing back over the divide and returning to travel writing. Rabinow makes the synthesis stage of his analytical work into a first-person narrative and thus presents a novel approach that will reach a larger public than his usual discourse group. It is worth looking at how he handles interviews, especially since it is his key data collection method and also a method that travel writers may use if they have sufficient time in one place. Travel writers might be on a slow residency or be locals themselves and this gives them time for longer interviews that can be followed up less formally to clarify any missed points after a session of journaling with the original interview. Voice recording used to be very invasive in semi-structured interviews, but the ubiquity of the smartphone with its recorder makes case participants less self-conscious of the recording act. Letting others speak in the travel text is an ethical choice by the researcher and writer. Including findings from interviews also, as Rabinow demonstrates below, gives value to the text for readers interested in this place. Here, Rabinow recounts his repeated interviews with Richard, the owner and front of house worker in the Olive Grove hotel and bar on the road into Sefrou:

> The second day in Sefrou he [Richard] told me his life story. He was from an upper-middle-class Parisian family. He had left home in 1950 to seek adventure, ending up in Morocco [...] The lack of the usual French reserve and hostility was startlingly indicative, I mused, either of a transformation of French culture once it left France or an intense loneliness on Richard's part.
>
> *(Rabinow, 2007, 13)*

The first point to notice is the absence of direct speech, reproduced in inverted commas or speech marks. Almost the whole of Rabinow's account of 160 pages is without direct speech from the locals that he interviews. The short quotation above shows that it is often for economy that Rabinow paraphrases long sections in order to offer analytical knowledge based on his processing of longer periods of interview. In fact, he very clearly indicates this processing with the short interjection, 'I mused' (Rabinow, 2007, 13). He thus retains an ethical honesty in his reported speech of Richard and further begins to sow the seeds of two themes that readers will come to realise are part of the synthesis of new knowledge from this fieldwork. These are the disconnection of the French immigrants to Morocco in the early 1950s from the other two waves of incoming French residents. Hence Richard's loneliness. The second is the hint that cultures can change when transplanted into a different place. Rabinow reminds his readers that it was almost accidental that he had captured these interviews:

> The structural possibilities of the situation were also ideal for collecting information [Rabinow often spoke with Richard for long periods in a relaxed manner, and neither dominated the relationship]. I did not conceptualize it in this way at the beginning, and for this reason I never systematically pursued this situation.
>
> *(Rabinow, 2007, 18)*

It is a valuable lesson that writer-researchers should remain open to these encounters even if unplanned in their itinerary. It demands, too, that writers find ways of capturing the essence of the interview as close in time as possible to the chance conversation. In a work much later than his Morocco book from the late 1970s, Rabinow turns the whole issue of reporting speech from the field in a discussion of Thucydides' *Peloponnesian War* (Rabinow, 2008, 66–72). Thucydides famously structures his account of the war around 27 speeches, which he places at turning points in the politics of the conflict. He writes them as if verbatim, but does explain earlier that he found it difficult to remember exactly what was said in terms of what was spoken. However, Thucydides maintains that he is giving a true sense of what was said by the protagonists. Rabinow says that this reporting method is no longer used in history writing, nor in journalism nor even in his own social science discipline (Rabinow, 2008). Travel journalism, though, often does include a quotation from a local or a company spokesperson in order to reinforce a point by the travel journalist but long speeches by company representatives are not reproduced verbatim. The questions to pose to the way the other's spoken word is reported are, first, how well they have been represented and further whether the writer's interpretation of their contribution is fair. For the writer, too, the value question must be considered; will these words from the local contribute to the synthesis of the research findings and contribute to the affect created by the literary text?

# 2

# AFFECT, EXPERIENCE AND THE LITERARY TEXT

## What is a literary text?

The travel writing explored in this chapter is subjective and literary. But what does literary mean, when used to describe discourse in the tourism industry? And how does it differ from non-literary writing? What characteristics give a text the so-called literariness? Kuijpers and Hakemulder (2018, 620) mention 'the unique qualities of literary style'. Dixon et al. (1993) argue that literariness is, in fact, an effect that happens 'during the interaction between reader and identifiable text qualities' (Kuijpers & Hakemulder, 2018, 620) and mention the poetic aspects of the text (Kuijpers & Hakemulder, 2018, 621). In reading a literary text, the reward is different from the reward in reading a non-literary text (i.e. mathematical, legal, philosophical) (Kuijpers & Hakemulder, 2018). Very often, emotions that are triggered when reading a text are among the key characteristics of literariness. Of course, comprehension of texts encompasses emotional and cognitive aspects (Kneepkens & Zwaan, 1994), and also reading a legal text can trigger emotions.

A particular type of travel writing is related closely to emotions, feelings, and covers a spectrum from hopes to disappointments. People read for various reasons: To learn, to enjoy, to escape from reality and to relax (Kneepkens & Zwaan, 1994). According to Kneepkens and Zwaan (1994, 125), literary texts 'are among the few text genres that may arouse emotions in the reader'; cognitive processing, however, precedes the emotional experience, but the latter may influence the cognitive processing. Miall and Kuiken (1998) are also among researchers who emphasise feelings in literary texts. They (Miall & Kuiken, 1998) argue that arousing feelings in the reader may contribute to longer reading times of the texts.

Miall (1988, 259) states that '[l]iterary stories are generally about people in social situations'. Forgas (1981) emphasises the significance of affect in

DOI: 10.4324/9781003178781-3

modelling social situations. As a source of affects and attitudes in person perception and memory, Figurski (1987) and Miall (1986, 1988) point out the key role of the self. Miall (1988, 260) states: 'If understanding of the self and perception of social episodes involve a central role for affect, it seems likely that affect also plays a significant part in response to narrative'. In reading and interpretation, feelings are central (Miall & Kuiken, 2002).

There is no doubt that stories are intended to evoke various affective responses (Brewer & Lichtenstein, 1982; Miall, 1988). Miall (1988, 260), in fact, proposes that 'affect plays the primary role in understanding literary stories' and is also central to reader response. He (Miall, 1988, 260–261) writes:

> Comprehension involves relating the text to existing knowledge of the world, which can be described as codes, frames, schemata, etc.; but there is also the sequential, experiential aspect of reading which uncovers ambiguity, indeterminacy, and conflict between schemata, and these require the reader's interpretative activity, during which schemata are shifted, transformed, or superseded.

Kneepkens and Zwaan (1994) mention the role of language in arousing emotions – the authors use the term 'seductive details' (Kneepkens & Zwaan, 1994, 128) to refer to 'information that was emotional, vivid, dramatic, exciting and personally involving, like facts about the personal life of the protagonist'. Nell (1988) points out that readers may devote their attention to a text unconsciously, and thus seemingly automatically or by applying themselves consciously and subjectively. Of course, in academic reading, readers will approach the texts hermeneutically, endeavouring to interpret or critically respond to the writing.

Miall and Kuiken (2002, 223) assert that in literary response,

> feelings can be sorted roughly into four domains: (1) Evaluative feelings toward the text, such as the overall enjoyment, pleasure, or satisfaction of reading a short story; (2) Narrative feelings toward specific aspects of the fictional event sequence, such as empathy with a character or resonance with the mood of a setting; (3) Aesthetic feelings in response to the formal (generic, narrative, or stylistic) components of a text, such as being struck by an apt metaphor; and (4) Self-modifying feelings that restructure the reader's understanding of the textual narrative and, simultaneously, the reader's sense of self.

Another significant observation by Kneepkens and Zwaan (1994, 128) is that 'attention given to interesting information requires fewer resources than attention given to uninteresting information'.

In their research, Miall and Kuiken (2002) pointed out a significant conclusion, namely, that readers react very individually to the same texts. So far, many studies have confirmed relation between reading and empathy

(Johnson, 2012; Johnson, 2013; Kidd & Castano, 2013), and reading and re-flection (Miall & Kuiken, 2002; Sikora et al., 2010; Koopman, 2016).

Koopman (2015) also investigated the effects of text genre, personal factors and affective responses during reading on two types of empathy: Empathic understanding and pro-social behaviour (donating), and the results of the research pointed to the relevance of a reader's personal characteristics.

All the above arguments support the idea that writing and reading needs to be studied as an essential part of tourism discourse, destination branding and marketing and, thus, an essential part of tourism education. The inclusion of literary reading and creative writing in education at tertiary levels is significant for many reasons – in travel, tourism and heritage interpretation. According to Fialho and Kuzmicova (2019), literature is important in education and at the workplace – in education for broadening students' cultural, social and personal horizons, and in the workplace, literature can promote interpersonal competencies, social success and moral enhancement. Reading can produce better social-cognitive abilities (Mumper & Gerrig, 2017) and contribute to empathy (Thexton et al., 2019). Reading literature, in fact, promotes well-being, healthy development and transferable learning (Darling-Hammond et al., 2020), since it often contains passages that require certain intellectual application and sensibilities to comprehend. Novels from the genre of Romanticism, for example, demand a careful reading, and sometime re-reading to grasp the concepts, as the example from Jane Austen shows below.

### Heteroglossia and free indirect discourse

A feature of literariness is the inclusion in the text of different voices or discourses, not simply the narrator recounting what happened but also the speech of others, transcribed as faithfully as possible, or reported as indirect speech. A third type of discourse is called, free indirect discourse (Russell, 2010). Adam Russell presents and analyses an example of FID from Jane Austen's novel, *Persuasion* (1818). Here is a detailed section of the passage:

> "Is not this song worth staying for?" said Anne, suddenly struck by an idea which made her yet more anxious to be encouraging.
>
> "No!" he replied impressively, "there is nothing worth my staying for;" and he was gone directly.
>
> Jealousy of Mr. Elliot! It was the only intelligible motive. Captain Wentworth jealous of her affection! Could she have believed it a week ago—three hours ago! For a moment the gratification was exquisite. But alas! there were very different thoughts to succeed. How was such jealously to be quieted? How was the truth to reach him? How, in all the peculiar disadvantages of their respective situations, would he ever learn her real sentiments? It was misery to think of Mr. Elliot's attentions.—Their evil was incalculable.
>
> *(Austen, 1818, 190–191)*

The final paragraph, especially from the words, 'But alas!' is FID, detaching itself from the direct thoughts of Anne, whilst still retaining, initially at least, her manner of speaking. FID, too, is shifted away from a text that is clearly spoken by the narrator. The section of FID can often be the moral voice, or doxa, of the society surrounding the text at the time of its writing. Moralistic interjections, for example, 'That would never do' fit this category of FID. This second example, taken from *Mansfield Park*, shows this social doxa at work:

> It was the abode of noise, disorder, and impropriety. Nobody was in their right place, nothing was done as it ought to be. She could not respect her parents, as she had hoped.
>
> *(Austen, 1814, Ch 3)*

The narrator's voice is only re-asserted with the use of the third person pronoun: 'She could not respect', so those first two sentences are free, that is, neither the narrator nor the character is speaking. When reading for pleasure, this depth of hermeneutic analysis is not consciously performed. The socialisation then is taken as a whole by readers and allows for a moral, or an emotional or ironic stance to be constructed by the writer towards events or surrounding spatial practices. For example, in a tourism setting, an FID interjection might be 'No-one would sun-bathe beside the hotel pool!'

When holidaymakers first arrive in the destination resort, they are unaware of what is permissible in this social space; Pierre Bourdieu defines the limits of the permissible as the *doxa* in sociological terms:

> Every established order tends to produce the naturalization of its own arbitrariness [...] out of which arises the *sense of limits*, commonly called the *sense of reality* [...] which is the basis of the most ineradicable adherence to the established order. [...] This experience we shall call *doxa*.
>
> *(Bourdieu, 1977, 164)*

In the literary text, it is free indirect discourse that speaks with the authority of *doxa*, that is, the FID says what can be practised here, it thus socialises by moulding the thoughts, leaving the created literary world limited but comprehensible, and because of this, realistic. In the same way that the holidaymaker is faced with a blank social canvas when arriving in a new town, the reader of the novel is initially free to interpret any utterance in any way, social mediation begins to limit the practices that may be played out until the reader-tourist adheres to the established order. In the novel, this social control is performed by free indirect discourse. The literary travel writer adheres to this to maintain a realism in the travel text, otherwise the readers will cease to believe the recounted journey.

## *Interpellation or calling into the text*

Catherine Belsey (1980) develops from Althusser's idea of interpellation (Althusser, 1971), the proposition that the reader is hailed by or called into the literary text during reading. At a direct level, an example of this can be seen in Modiano's novel in its more recent English translation by John Cullen, *Villa Triste* (2016). This novel is used an example of literary place-making because it offers up street names and nostalgically talks of hospitality locations in the holiday destination of Annecy:

> the rotunda-shaped café next door has also disappeared. Was it called the Dials Café, or maybe the Café of the Future?
>
> *(Modiano, 2016, 3)*

Here, the narrator poses questions, which although they may only be rhetorical, also have the agency of addressing the reader. Then, shortly after that, he continues:

> You pass shop windows – the Chez Clément Marot bookstore [...] you can see the red and green neon lights of the Cintra shining at the end on your left.
>
> *(Modiano, 2016, 4)*

The reader is now directly addressed in the second person, as 'you'. And furthermore, the deictic phrase 'on your left', places the reader in the urban space. The holiday town of Annecy has been co-created with the reader by the literary language devices of deixis and interpellation.

## The aesthetics of urban space in tourism

Aesthetics is vital to the 'human sense of well-being' and industries, like tourism and heritage, involved in 'catering to aesthetic satisfactions' (Porteous, 1996, 5). It seems that the beautiful represents the essence of tourism and tourism communication, which has always also involved aesthetics, which is, according to Prall (1929, 45), basic to human nature. According to Di et al. (2010), aesthetic values are at the centre of human perception of a tourism destination. Aesthetic value is also one of the significant criteria in the evaluation of the application process for natural areas to be designated as World Natural Heritage Sites by UNESCO (Di et al., 2010, 59). In the first half of the 1980s, Zube et al. (1982) dealt with landscape perception and, in recent decades, especially since the 1980s, the growth of the tourist industry has led leaders and politicians to reconsider landscapes as revenue generators (Porteous, 1996, 10). Aesthetics explores the nature of beauty and comprises one of the five classical fields of philosophical inquiry, together with Epistemology, Ethics, Logic and Metaphysics (Sporre, 2006, 7), and is often discussed in tourism literature (Knudsen & Greer, 2011; Austin, 2007;

Scarles, 2004). To create a valuable experience for tourists and to present the built heritage of destinations as beautiful is one of the goals of tourism advertising and, consequently, the aesthetic dimension of an attraction or a destination is significant, despite the fact that the term aesthetics is highly disputed in philosophy (Todd, 2012, 65). Knudsen et al. (2015, 179) speak of three threads of aesthetics in tourism: (1) Tourism from the point of view of semiotics; (2) Foucauldian discourse analysis as an ocularcentric activity; and (3) Anthropologic origin which states that tourism has much in common with ritual performance.

The aesthetic characteristics of an attraction or a destination influence the experiences and satisfaction of tourists and contribute to their wish to return to places and monuments in the town. According to Alegre and Garau (2010), a destination's aesthetic characteristics have been an essential element of many perception and satisfaction image scales used in tourism research. Western philosophy and modern sociology still ask whether beauty is a property of the artefact or the person contemplating the object or townscape. Turning to the question of tourists' experience, which is 'a critical concept in tourism marketing and management literature' (Kirillova et al., 2014, 282), tourism aesthetics possesses its own characteristics because 'the tourism experience involves the full immersion of an individual into an environment that may be distinct from his/her everyday living surroundings' (Volo, 2009 in Kirillova et al., 2014, 283). Whether tourists perceive an attraction as beautiful could be related to their previously experienced environments (Maitland & Smith, 2009).

It should be observed, however, that in appreciation of the observed landscape, 'what is aesthetically relevant is knowledge of why it is, what it is, and what it is like, whether or not that knowledge is, strictly speaking, scientific' (Carlson, 2002, 549). Thus, according to Carlson (2002, 549), who speaks of the 'aesthetic relevance of information', information about an observed object's histories, functions, its roles in our lives, is crucial; it is tour guides and travel writers who provide that significant information. Consequently, much is dependent upon how the presentations of objects and attractions are recounted by writers, reviewers or guides. In fact, the travel writer's information plays a central role in the perception of a townscape or museum artefact. 'The aesthetic relevance of such information seems especially evident for environments that constitute important places in the histories and cultures of particular peoples' (Carlson, 2002, 550). What is important is 'an emotionally and cognitively rich engagement with a cultural artifact, intentionally created by a designing intellect, informed by both art-historical traditions and art-critical practices, and deeply embedded in a complex, many-faceted art world' and 'emotionally and cognitively rich engagement with an environment, created by natural and cultural forces, informed by both scientific knowledge and cultural traditions' (Carlson, 2002, 551). According to Ittelson (1978), tourism aesthetics involves multi-sensory experiences, which may incorporate many relations besides the one between a tourist and the environment.

## Tourists' experiences and user-experience design (UXD)

Cultural experiences are, according to Richards et al. (2020, 1), 'one of the most important elements of tourism production and consumption'. Tourists planning a visit to a city or to a specific tourism space may consider what experience they hope to enjoy there. It is a question that managers of these spaces and designers of audience experience ask and return to regularly. Tourists' experiences may be planned as self-guided routes (by providing maps, brochures, mobile apps) or as guided tours with guides providing interpretations on site.

McIntyre (2009) reminds researchers and designers of this question applied to museums and galleries. His research shows that a visitor experience has a range of dimensions when reported upon, from cultural bathing when near the artefacts through to spaces that 'can allow respite from, or detached contemplation of, object and experience compression by, for instance, reading museum literature' (McIntyre, 2009, 167). McIntyre's research suggests that these spaces of different experiences need signposting by the museum's leaflets and plans so that visitors can make informed choices about the experience they are likely to enjoy. The researching travel writer can record their own emotions and experience in such a way as to provide this planning literature for the visitors, and in this way contribute to unique cultural experiences. This text can be, rather than just a direction indicator, a re-design of the experience by the integration of a time element into the use of the space. McIntyre's (2009) work also includes the constant challenge of tourism between whether the tourists are learning from authentic artefacts and local practices and whether they are seeking entertainment. It is a question that stands over cultural artefacts and intangible cultural heritage at every moment in tourism history from whether the object was plundered at some time in the past or sold for the benefit of a local trade, or created and curated by makers in the culture that is presenting them to a contemporary consuming group. This is exactly why the locals, who are recognised as the bearers of cultural heritage and culture, have a crucial role in the management of cultural heritage, its preservation, presentation, and promotion, and need to be actively involved in the creation of various tourism activities, offers and strategies.

Emotions are closely related to tourists' experiences and in consumption experience emotions are vital, but under-researched (Richins, 1997). Studies have shown that emotions are significant in choosing and using a certain product, and they impact the satisfaction and selection of the next product or destination (Tung & Ritchie, 2011). In another study, Nawijn et al. (2013) were researching tourists' feelings on vacations. According to their research (Nawijn et al., 2013), it is difficult to measure emotions since they are complex and short-lived, and in addition to that individuals have many emotional experiences within a single day. Many scales for measuring emotions have been developed, but for the purpose of developing tourism experiences and products, the Destination Emotion Scale developed by Hosany and Gilbert in 2010 that focuses on the factors joy, love, and positive surprise (Ilić et al., 2021). Nawijn et al. (2013) also point out that the recall of emotions over a period of several days or weeks (for example after ending the holidays) is not considered valid, and they argue that emotions must be measured close to the moment in which they occur. This is valuable research information for travel writers. If the goal of a travel writer is to capture feelings, travel writing should be done during the trip and not after it trying to recall feelings and memories. However, since an effective method for measuring emotions is experience, travel writing has the potential to be a valuable tool not only for branding destinations, products and services but also for sampling tourism experiences and emotions, and further on, measuring tourist satisfaction.

> By making use of their lived bodies, individuals are capable of engaging in a particular sort of imaginative play through which memories of the past, the latent reality and the actualized perceived present are conjured together, informing one another.
>
> *(Tursić, 2019, 211)*

Tursić's (2019) view here is that managers of tourism spaces need to be aware that their diverse visitors inhabit an imaginative and topological space conditioned by their accumulated knowledge. The literary travel writer, then, must be simultaneously aware of larger cultural narratives but at the same time be prepared to create new, supplementary narratives to add to the places being explored. One of the prime methods for this is to stimulate and then report upon the writer's own experiences in that space.

### The literary travel writer as cicerone and docent

A cicerone draws out interesting factual connections for tourists as they pass by the built heritage on their guided visits. The term comes from the name of Marcus Tullius Cicero (106 BCE–43 BCE), who was a prose writer and orator. Finding a balance between too much information and subtly pointing out why a technical point is interesting is part of the challenge of guiding the audience. The cicerone draws on many subject disciplines, the full range is discussed in the opening introduction to this book, to open up the beauty or intended meaning of the architect, builder or designer. This will enrich the experiences of the visitors.

Take, for example, the built heritage in London of the St Pancras International station from its 2007 refit as the Eurostar train terminal. It is an architectural space where travellers sit to wait before going to the upper floor for their trains to Paris, when they are called. It is an experience enjoyed by many, since Eurostar carries 11 million passengers a year. It is a space then, where many tourists will be reading guide books, holiday reading or travel literature to pass the time enjoyably or to fine-tune their vacation itineraries. Some will be racing through the texts they have, while others will be savouring the literary space created by the text. The waiting room space can be a metaphor for these two approaches to reading: business-like or enjoying the aesthetic construction.

To better appreciate and thus enjoy the aesthetics of the built surroundings, the skilful cicerone would add some factual information that draws attention to what might at first seem mundane, or even invisible, the floor. Six thousand square meters of solid wood Jatoba were machined by Atkinson & Kirby in Ormskirk to be laid in time for the opening on Tuesday 6 November 2007. Back then, it was the largest single space of wooden flooring in Europe. Jatoba is a hard wood from accredited forests in Brazil. When first laid, it has a pink to tan tint but darkens over time to a deep, vibrant red with black striping. Which is why if you return for another holiday in Paris in three or four years' time, the flooring will be a much richer colour. The true colour the designers had planned requires patience. To enjoy this aspect of the space, the visitor requires patience.

A tour guide could recount that fact in a single plateau or stop-and-tell session; however, reading this in a piece of travel literature demands technical engagement from readers. It would take 300 words to communicate all the aspects that create its value and hence would be sufficient architectural digression from the travel story for this building and no further element of design would be mentioned. The same principle applies to museum visits within a travel story. The details of one painting or sculpture is enough to inform the readers without the text becoming a museum guide book or the equivalent of the museum's docent taking a group around the collection. In written practice, the economy of heading straight to one artefact or one viewing point unlocks the readers' sufficient knowledge without distracting from the theme of the travel story. Indeed, Gide says that the literary text should be stripped of all that does not specifically belong to the story (Gide, 1925, 84), in which case the literary travel writer should ensure that the highlighted artefact or view does not distract from the theme or story. The literary travel writer, Vita Sackville-West achieves this careful economy in her exploration of the Persian city of Isfahan, in the section below.

## Looking down, looking through, mystery in storytelling

When the literary writer presents a scene, a person, a group, and even an artefact which cannot be seen clearly, mystery is added. Rather than annoying readers of this type of text, it leaves an unresolved space in which they can co-create, rather like a clue in detective fiction. A partially obscured view through trellis-work in a garden can open the field for readers to participate more in resolving the mystery. Sackville-West (2007) presents her readers with a skilful paragraph that draws a map of the key sites for visitors to Persian Isfahan. She climbs up onto the viewing platform of the Safavid Ali Qapu Palace which opens eastwards down onto Naqsh-e Jahan Square but behind, to its west lies the Chehel Sotoon pavilion with its long reflecting pool. Note that there are some slight spelling changes since her writing was published.

> Ali Carpi trembled under my feet; at first I thought that an earthquake was about to level Isfahan in one magnificent cloud of dust, but soon realised that the trembling was due merely to the insecurity of the frail old building. So I remained where I was. I could look down into the tank of the Hall of Forty Columns – the Chel Setun – fringed with umbrella pines and ending squarely at the foot of the little palace. I could look down into the Meidan, where the tiny figures strolled, or a carriage like a toy crossed the square followed by a swirl of dust. I could look almost into the courtyard of the mosque – that sanctuary forbidden to the unbeliever. To look down upon a city from a roof high above its roofs is to gain a new aspect; everything appears at an odd angle, and freakish framings make little complete pictures like the vignettes in medieval paintings; thus between the blue domes I got a group of brown houses, with the profile of the hills behind; or through the ogives of a window I got one rounded bubble of blue dome, like a huge mappemonde, the continents and seas represented by the stains where the tiles had fallen off. There was plenty to amuse an idler on the roof, and to descend from those airy solitudes to the earth below was like coming down into a world of which one had taken wily advantage, gained a surreptitious and almost dishonourable acquaintance.
>
> *(Sackville-West, 2007, 112)*

The proximity of the reflecting pool of the Forty Columns pavilion is not realised by visitors in this century because today the focus is on looking down onto the vast square although the Ali Qapu floor still trembles a hundred years after Sackville-West made her notes there in the early 1920s. The framing and point-of-view that she communicates lifts her readers over that urban space, letting them hover over the bustling city. However, the set piece is not a distraction from the story, she is true to Gide's advice. First, she cannot quite see into the forbidden space of the courtyard, reminding readers of the local culture. Moreover, she is carefully building this chapter's larger theme of her relationship with the Persian city. It is a strong emotional feeling, one borne out of physicality as much as memory. Notice how the paragraph opens both with movement, the trembling building, and the immediacy of a sensual experience which was felt from beneath rather than merely seen.

Walter Benjamin, in his analysis of how stories work, proposes that 'at least half of the art of storytelling consists in keeping one's tale free of explanation' (Benjamin, 2019, 35). Unlike information, the artfully created story is not used up because it always leaves this gap for its readers to complete as they explain the meaning to themselves. They must draw the conclusions and hence co-create, often with different interpretations at different times, especially if the story is re-read. Modiano presents a staged unfolding of the narrative function of a local mineral water in his place-making story from 1989 *Vestiaire de l'enfance, Childhood Locker*. Readers first encounter the odd taste of the mineral water as part of the character sketch of a time-served expat, who is inseparable from this liquid, which is as heavy as mercury. From this introduction to the water, the story keeps it as an important signifier in the Rosal Café, where it is served in tiny carafes. By now, its association is that tourists do not drink this water from the port-town's hinterland. So, when the narrator sees it being carried to a new customer, he is naturally interested in who it might be. Here, Modiano uses the mystery of looking through to an obscured scene; the grille of an ornate iron gate and the contrast of the bright sunlight make the client difficult to see. This screening is particularly powerful here because the expression on the face of a someone who has tasted this water for the first time is unmistakably revealing about their status in the seaside port, are they passing through or almost locals? The whole novel, as a Benjaminian story is not used up, as one Goodreads reviewer puts it: 'I expected more and I didn't get the answers I was looking for' (Carla, 2015).

## The DRAMMA model applied to travel literature

Travel writing narratives, according to Laing and Frost (2017, 110), have the ability to shape the perception about tourist places, and are 'a useful tool for understanding tourist experiences'. These narratives have many functions; they may inspire, encourage, educate, inform, entertain, etc. and, thus, contribute to subjective well-being of writers and readers. Some recent studies have researched the relationship between travel writing and well-being (Newman et al., 2014; Laing & Frost, 2017; Kujanpää et al., 2021), analysing the so-called DRAMMA model that includes six psychological needs: detachment, relaxation, autonomy, mastery, meaning and affiliation.

Newman et al. (2014), who developed the DRAMMA model, explain that detachment is usually associated with detachment from work or to detachment from troubling elements of an individual's life; recovery is associated to rest and recuperation; autonomy to engaging into new activities that require independence and self-direction; mastery is referred to involvement into activities that allow individuals to develop and grow; meaning is associated to finding some purpose and value in the individual's life and affiliation is associated to relationships and connecting to others and the community.

These psychological needs cover both eudaimonic and hedonic dimensions of well-being (Laing & Frost, 2017) and connect leisure to well-being and optimal functioning (Kujanpää et al., 2021). Common elements of eudaimonia are growth, authenticity, meaning and excellence, while hedonic are enjoyment, pleasure and absence of discomfort (Huta & Waterman, 2014, 1427).

Travel writing is a ticket to well-being, especially through processes of self-discovery (Laing & Frost, 2017) and identity development. By employing the DRAMMA model (Newman et al., 2014), and the psychological needs of detachment, relaxation, autonomy, mastery, meaning and affiliation, travel writing tours can be further developed into tourism products

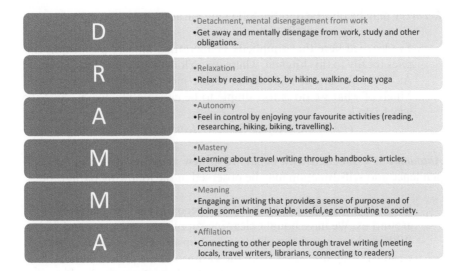

**FIGURE 2.1** The DRAMMA model. Adapted by Mansfield & Potočnik Topler after Newman et al. (2014).

for various segments of tourists and, thus, contribute to their emotional and physical well-being (Figure 2.1).

All the elements of the DRAMMA model (detachment, recovery, autonomy, mastery, meaning, affiliation) can be found in the novel by Frances Mayes (1996) *Under the Tuscan Sun*. On the levels of sentences and paragraphs, many examples can be observed; thus, below key examples have been abstracted as a guide to further work:

***Detachment*** which is usually associated with detachment from work or detachment from troubling elements of an individual's life, may be observed in the following paragraph:

> I AM ABOUT TO BUY A HOUSE IN A FOREIGN country. A house with the beautiful name of Bramasole. It is tall, square, and apricot-colored with faded green shutters, ancient tile roof, and an iron balcony on the second level, where ladies might have sat with their fans to watch some spectacle below. But below, overgrown briars, tangles of roses, and knee-high weeds run rampant. The balcony faces southeast, looking into a deep valley, then into the Tuscan Apennines. When it rains or when the light changes, the facade of the house turns gold, sienna, ocher; a previous scarlet paint job seeps through in rosy spots like a box of crayons left to melt in the sun. In places where the stucco has fallen away, rugged stone shows what the exterior once was. The house rises above a strada bianca, a road white with pebbles, on a terraced slab of hillside covered with fruit and olive trees. Bramasole: from bramare, to yearn for, and sole, sun: something that yearns for the sun, and yes, I do.
>
> *(From the Chapter Bramare, [no pagination])*

*Recovery* may be observed in the following paragraph:

> I've learned here that simplicity is liberating. Simca's philosophy applies totally to this kitchen, where we no longer measure, but just cook. As all cooks know, ingredients of the moment are the best guides. Much of what we do is too simple to be called a recipe—it's just the way to do it.
> *(From the Chapter Summer Kitchen Notes, [no pagination])*

*Autonomy* is associated with engaging in new activities that require independence and self-direction may be observed in the following paragraph:

> Every day we haul and scrub. We are becoming as parched as the hills around us. We have bought cleaning supplies, a new stove and fridge. With sawhorses and two planks we set up a kitchen counter. Although we must bring hot water from the bathroom in a plastic laundry pan, we have a surprisingly manageable kitchen. As one who has used Williams-Sonoma as a toy store for years, I begin to get back to an elementary sense of the kitchen. Three wooden spoons, two for the salad, one for stirring. A sauté pan, bread knife, cutting knife, cheese grater, pasta pot, baking dish, and stove-top espresso pot. We brought over some old picnic silverware and bought a few glasses and plates. Those first pastas are divine. After long work, we eat everything in sight then tumble like field hands into bed. Our favorite is spaghetti with an easy sauce made from diced pancetta, unsmoked bacon, quickly browned, then stirred into cream and chopped wild arugula (called ruchetta locally), easily available in our driveway and along the stone walls. We grate parmigiano on top and eat huge mounds. Besides the best salad of all, those amazing tomatoes sliced thickly and served with chopped basil and mozzarella, we learn to make Tuscan white beans with sage and olive oil. I shell and simmer the beans in the morning, then let them come to room temperature before dousing them with the oil. We consume an astonishing number of black olives.
> *(From the Chapter titled A HOUSE AND THE LAND IT TAKES*
> *TWO OXEN TWO DAYS TO PLOW, [no pagination])*

The following paragraph is an example of *mastery*, that in the DRAMMA model, refers to involvement in activities that allow individuals to develop and grow:

> Now I love the quick Mass in tiny upper Cortona churches, where the same sounds have provided a still point for the residents for almost eight hundred years. When a black Labrador wandered in, the priest interrupted his holy spiel to shout, "For the love of God, somebody get that dog out of here." If I stop in on a weekday morning, I sit there

alone, enjoying the country Baroque. I think: Here I am. I love the parade of relics through the streets, with gold-robed priests travelling along in a billow of incense, their way prepared by children in white, scattering the streets with petals of broom, rose, daisy. In the noon heat, I almost hallucinate. What's in the gold box held aloft with banners—a splinter from the cradle? Never mind we thought Jesus was born in a lowly manger; this is the splinter of the true cradle. Or am I confused? It's a splinter of the true cross. It is on its way through the streets, brought out into the air one day a year. And suddenly I think, What did that hymn mean, cleft for me, rising years ago, perpendicular from the white board church in Georgia?

*(From the Chapter titled RELICS OF SUMMER, [no pagination])*

*Meaning* is associated with finding some purpose and value in the individual's life:

Living here, I've intensely reconnected with nature. The land, we've learned, is always in a state of lively evolution. The lane of cypresses and lavender we planted is beginning to look as though it has always been there. The slender cypresses, just my height when we planted them, now look like those exclamation points we see punctuating the Tuscan landscape. Between them, the lavender's amethystine radiance lights the path. Roses, marguerites, lavender, pale yellow petunias, and lilies on our front terraces have made the ivy and blackberry jungles just a memory. The biggest change is grass. Grass is not Tuscan. We lived with a mown and watered weed lawn for several years. Lovely in spring and early summer, it looked forlorn in August. No amount of precious water kept it alive. One September week, with the help of three neighbors, we unrolled miles of sod trucked from Rome. The irrigation system looks like the Chicago Fire Department's command central. Neither of us understands it completely. Now, a few years later, the clovers and tiny flowers have staged a comeback—grass giving over to weed again.

*(From the Chapter titled BEN TORNATI*
*(WELCOME BACK), [no pagination])*

*Affiliation* is shown in the relationships and by connecting to others and the community:

ON OUR FIRST MORNING BACK IN CORTONA, after several months in California, my husband Ed and I walk into town for groceries. First, I drop off film to be developed at Giorgio and Lina's photo shop. "Ben tornati," Giorgio shouts, welcome back. Lina comes from behind the counter and all four of us exchange the ritual cheek kisses. Finally, I've learned to go to the right, then left, thereby avoiding head swivels or full-lip encounters. Lina wastes no time. In the confusion of other customers and the small space, I piece together, "We must go for dinner," "In the country, but close," and the ultimate praise, "She cooks like my mother." Giorgio interrupts. "Saturday or Sunday? I prefer Saturday but would make the supreme sacrifice." He looks like an older, more mischievous version of Caravaggio's Bacchus. He's the town photographer, present at every wedding and festival, and is known to like dancing. Last summer we shared an all-goose feast with him and Lina—and, of course, about twenty others. Every celebration involves an infinitely expandable table. "The pasta with duck." He shakes his head. "That duck squawked in the pen in the morning and came to the table at night." "What's the sacrifice?" Ed asks. "Soccer in Rome." "Then we'll go Saturday." Ed knows soccer is sacred. We cross the piazza and run into Alessandra. "Let's go for coffee," she says, sweeping us into the bar to catch up on news. She is newly pregnant and wants to discuss names. As we leave her and head toward the grocery store, we see Cecilia with her English husband and two magical little girls, Carlotta and Camilla. "Dinner," they say. "Come when you can. Any night."

*(From the Chapter titled BEN TORNATI (WELCOME BACK), [no pagination])*

## Travel writing as a catalyst for the toureme

The toureme is the creation of value during a holiday. This making of meaning takes place as a focus of holidaymakers' cultural capital particularly from their reading, formal education or hobby interest, plus an experience they are enjoying in the tourism space, and the third component, a moment of recounting, for example, writing out a postcard or a text to a friend. Or, indeed, keeping a journal of the experience. The concept of the toureme draws from Bourdieu's cultural capital being enacted at the scene of leisure, since the tourists have an emotional opportunity to feel their knowledge vindicated. For example, in a novel they are reading, they understand that their seaside town in Brittany was also a fishing port; they encounter a disused sardine canning factory with its chimney, and their knowledge from the novel is echoed in the urban space they are enjoying. To fully realise this moment as a toureme, though, some process of calling on the holidaymakers' imagination is required, and this last element of the toureme is the recounting for a friend through the act of writing to tell of their discovery. The theory behind the re-telling is developed by Kristin Ross (1988) from the points in the essay on the secret law of the tale or *récit* by Maurice Blanchot, from the original in French published in 1959:

> They enticed him to a place which he did not want to fall into and, hidden in the heart of The Odyssey, which had become their tomb,

they drew him – and many others – into that happy, unhappy voyage which is the voyage of the tale – of a song which is no longer immediate, but is narrated, and because of this made to seem harmless, an ode which has turned into an episode.

*Blanchot (1998, 445)*

The recounting or the *récit*, as Blanchot calls it, opposes the doxa of the realist novel, breaking out of social acculturation (Ross, 1988) of the dominant genre not to give a report on what happened but to be an actual place and a moment of created knowledge. Blanchot takes up the point, 'The *récit* is not the narration of an event, but that event itself' Blanchot (1998, 447) cited in (Ross, 1988, 49). If literary travel writers are aware of this phenomenon, then they can offer stories in which they recount their toureme moments, and if their cultural capital is shared by their readers, then those readers can seek out the places and experiences from the writers' stories. Travel writers, too, can include catalysts for these touremes by supplying enough information about the literary author who wrote at that spot or set an event in the novel in that place, to provide all the conditions for toureme value. This is, of course, an extremely sensitive undertaking since too much information can dispel any mystery leaving the tourists with no discovery to make from the clues themselves. Further, for the toureme to happen, the tourist must recount this in some way, for example, through a text message to a friend, so that the tourist's imagination has been exercised.

Tourists visiting Isfahan might be carrying Sackville-West's travel book with them when they climb the trembling steps of the Ali Qapu Palace and once there try to recreate her mysterious views of the Chehel Sotoon pavilion and reflect on what can still be seen and what has become even more built up, then record their findings for friends to create their own toureme.

# 3

# TRAVEL WRITING IN PLACE BRANDING

## The relationship between literary travel writing and destination value

Literary travel writing, which can be considered as engaging readers in the textual work for itself, is one of the tools for attracting tourists and visitors to destinations in the same way that traditional literary tourism has been employed (Sardo & Chaves, 2022). This engagement with the places in the literary text creates a value for those places in the readers who have contributed personal effort in working through the text. Through the narrative they have come to know the places. Thompson (2011) argues that travel writing's reputation rose sharply in the second half of the twentieth century, with a new generation of critically acclaimed authors. These are regarded as literary travel writers and include Paul Theroux, Bruce Chatwin, Ryszard Kapuscinski and Robyn Davidson. The phenomenon has been studied from the perspective of place branding and destination promotion, but the academic literature on more diverse topics dealing with travel writing is abundant. One of the reasons for studies across other disciplines of the phenomenon of travel writing is that the 'written text, which is both more private and more public than the spoken word, has its own unique powers of transforming reality' (Tuan, 1991, 690). However, Travel Studies as a research field only became consolidated in the 1990s (Culbert, 2018) and has had an increasingly important role in tourism management ever since. Culbert (2018, 343) characterises travel writing texts as 'diverse, hybrid, and wide-ranging' and points out that these texts 'have been invaluable to the study of our increasingly mobile and interconnected world', while, according to Alu and Hill (2018, 1), in travel writing texts, 'views and gazes express a narrative space from which narrator and reader scrutinise, judge and categorise the varied cultures and societies they explore'. Forsdick (2009, 287) argues that it 'is

DOI: 10.4324/9781003178781-4

an inherently transcultural, transnational, even translingual phenomenon', and the questions of cultural identities are, thus, significant, when it comes to travel writing (Rubiés & Bacon, 2000). Among famous travel writing texts were the descriptions of the Grand Tour: Francis Bacon's *Of Travel* (1612), Michel de Montaigne's *Journal de Voyage en Italie* (1580–1581), John Locke's *On Education* (1692) (Antosa, 2008) and of course the *Book of John Mandeville* and Marco Polo's *Devisement du monde* (Culbert, 2018). Even the first newspapers published travel accounts among their contents (Hutchins, 2013), and in the late eighteenth century, the travel book industry became very successful (Zilcosky, 2008). The answer to the question why this type of literature is well received by the audience might be in the fact that 'perhaps the most important function of language is the communication of narrative, conveying the actions of agents' (Yuan et al., 2018, 1298). By the beginning of the twenty-first century, the academic journal *Studies in Travel Writing* became an important source of literary analysis on travel writing as a finished cultural form, while *The Cambridge Companion to Travel Writing* as well as *The Routledge Companion to Travel Writing* became the key texts on for scholars interested in the socio-political issues surrounding the literary genre.

Travel writing is often discussed as a subtype of Literary Tourism (Busby & Klug, 2001; Gentile & Brown, 2015). The concept of Literary Tourism involves travel to a destination due to an interest in some form of literary association with that destination (Robinson & Anderson, 2002; Agarwal & Shaw, 2018) and is based on the belief that, by visiting literary sites, visitors are able to understand authors, their life and works more than by reading literary reviews (Potočnik Topler, 2016). Literary Tourism as a type of Cultural or Heritage Tourism is connected primarily to visiting 'both those places associated with writers in their real lives and those which provided settings for their novels' (Herbert, 1995, 33; Stiebel, 2007), is a phenomenon that has become an important niche in Tourism and it could be claimed that Literary Tourism comprises anything that is connected to literature, including literature events, performances and festivals. Smith (2012, 9) argues that book signings and creative writing courses are also parts of Literary Tourism. Her definition (*ibid.*) is, therefore, that Literary Tourism is 'a form of Cultural Tourism involving travel to places and events associated with writers, writers' works, literary depictions and the writing of creative literature'. She (2012, 11) also points out that, not only prose, drama and poetry inspire people to become literary tourists but also biographies and autobiographies. The phenomenon of literary tourism, therefore, complicates the issue of travel literature as an influence on tourists' behaviours. Authors and works not considered as belonging to the travel writing genre are being consumed as markers of place and encouraging visitors to locations.

When it comes to defining travel writing, many authors agree on the diversity of the genre (Holland & Huggan, 2000; Forsdick, 2009; Singer, 2016; Alu & Hill, 2018). Although Robinson (2004) explains extensively what is and what is not travel writing and argues that it is challenging to provide a definition of this phenomenon, he (2004, 303) describes it as 'writing about the experience of travel and visits to 'other' places'. Rubiés and Bacon (2000, 6), who state that travel literature is 'the genre of genres' and describe it as 'that varied body of writing which, whether its principal purpose is practical or fictional, takes travel as an essential condition for its production', also point out that travelling does not only involve the act of travelling but also as a rhetorical device within cultural discourses. Travelling created a rhetorical frame for various purposes, i.e. religious, satirical, epic. Monga (1996) emphasises several aspects of travelling and points out that travelling may also be a metaphor for life. Mansfield (2019) elaborates on the subject in his paper 'The role of travel writing practitioners in tourism management and place-branding research'. He (Mansfield, 2019) states that, initially, tourists or locals are provided with literary texts that have discoverable places. The next step is the led and self-directed movement of the tourist to the places of the novel. Due to its interdisciplinary nature, the definitions of place writing and travel writing are various and vague, but, what is significant, is that literature and text represent an essential medium in many disciplines.

For the positioning of destinations, communication and discourse are essential concepts (Potočnik Topler, 2018). In Tourism, travel writing is especially interesting as a tool for constructing the destination image and as a marketing tool (Woodside & Megehee, 2010; Séraphin, 2015). Place branding, which was originally considered in the frame of marketing, and only later as a part of destination management and tourism development (Van Assche et al., 2020), happens through languages; thus, linguistic choices are extremely important (Potočnik Topler, 2018). Place branding needs 'coordinated intervention' (Van Assche et al., 2020, 1275) and applying practices that correspond to a place's ambitions and objectives (Van Assche et al., 2020). In travel writing, traditional and contemporary literary, journalistic, promotional and tourism discourses merge. However, it is significant that travel writing remains credible by observing the following requirements: Obtaining reliable information, supporting evidence of what is stated in the text and statements by relevant people, for example interviews with the locals. When describing the journey or the route, Dann's (1996: 68) four sociological models of language usage for promotional texts in tourism may be employed: (1) the language of authentication (promotes the experience of the traveller as authentic, genuine, pure, opposing it to the banality of everyday life); (2) the language of differentiation (highlights the contrast between holiday and normal life); (3) the language of recreation (emphasises the recreational and hedonistic side of Tourism); and (4) the language of appropriation (tries to adopt an attitude of control and domination of what is unknown). In addition, commercially produced travelogues, as opposed to travel literature, need to be structured carefully, to be informative within limited space, to display a careful selection of writing techniques and make structural, lexical and grammatical choices that meet the expectations of the tourist reader. For this reason, they are centred around the tourist product in the destination, sometimes referred to as ground product in the tourist industry.

Alu and Hill (2018) explore the visual imagery in travel writings and in travel narrative in general, and argue that, in travel writing, 'views and gazes express a narrative space from which narrator and reader scrutinise, judge and categorise the varied cultures and societies they explore through writing and reading', while Bechmann Petersen (2006) speaks of the concept of cross media, and Scolari (2009, 587) of transmedia storytelling, which is 'a particular narrative structure that expands through both different languages (verbal, iconic, etc.) and media (cinema, comics, television, and video games.)'. Visual images (photos, receipts, sketches, maps, promotional materials) make travel texts authentic. 'For centuries, sketches, watercolours, engravings, lithographs, photography, film, and now digital media, have framed and recorded every aspect of our movements and experiences of dislocation' (Alu & Hill, 2018, 1). Of course, as Alu and Hill (2018, 11) observe, choice is significant in the travel writing process and 'not only of what to show, but also of whose photographs to include'. The literary travel writing of W G Sebald, for example, includes images of printed cultural artefacts collected almost as authenticating data along the route, which serve as clues for the avid reader to discover dates and build a chronology of the narrator's journey; see *Vertigo* (2016) in particular.

While travel journalism and promotional tourism texts often list the sites, travel options and ground product, literary travel writing is distinguished by its storytelling, which is 'a two-way interaction, written or oral, between someone telling a story and one or more listeners' (Sundin et al., 2018, x). In fact, literary travel writing is storytelling in writing, and storytelling has a long tradition. Its origins date back to folktales that are found in Sumerian, Egyptian, Chinese, Greek, Latin and Sanskrit (Hsu, 2008); Aristotle proposed the theoretical foundations of storytelling in *Poetics*, in which he established the so-called 'Aristotelian story structure' (Tierno, 2002, Preface ii). In *Poetics*, Aristotle (trans. Butcher, 2008) emphasises mimesis (imitation), catharsis and unity of action as essential characteristics and enumerates the following seven aspects of a good story: Plot, character, theme, diction, melody, décor and spectacle. People tell stories 'to entertain, to transfer knowledge between generations, to maintain cultural heritage, or to warn others of dangers' (Lugmayr et al., 2017, 15707). People enjoy listening to stories 'because they give them pleasure' (Smith, 2012, 24), prescribe expected behaviour, establish meaning and constitute identity (Godsil & Goodale, 2013). In Tourism, storytelling represents a specific approach to bringing information from the host to the guest, from the tourist guide to the visitor, which is characterised strongly by the [tourist] experience (Vitić-Ćetković et al., 2020). Products and destination are branded through stories as it was established that storytelling creates emotional connections between a destination and its target groups (Keskin et al., 2016; Ilić et al., 2021). By employing new technology and new media, stories – that are often characterised by high informational density (Bassano et al., 2019) – may contribute successfully to the distinctiveness of a tourism product or a tourism destination. This is significant, because businesses and employees reliant on selling ground product or on receiving income from visitors depend on tourists' engagement with the stories behind the products (Mossberg, 2007), and when it comes to applying storytelling in Tourism, it needs to be emphasised that

> the storytelling concept requires communication between different stakeholders: Tourism policy makers, destination organisations and service providers. It includes tourism organisations, public administration at local and regional levels, private partners, different types of service providers (hotels, restaurants, museums, shops etc.) and storytellers (individuals).
>
> *(Vitić-Ćetković et al., 2020, 93)*

The process of storytelling, according to Mossberg et al. (2010), is shown to have impacts on tourism stakeholders and the economic development of a destination in a cyclical manner; their work offers a useful diagram of this process at work (Figure 3.1).

It is not an exaggeration to argue that, nowadays, social media have drawn much of the investment for brand marketers' storytelling away from print newspapers, billboards and to some extent television and radio. In the West, the largest social media networks to date are Facebook (>2 billion users), Instagram (>1 billion users) and the main online video platform YouTube (1.5 billion users), founded in 2004, 2010 and 2005, respectively (van Laer et al., 2019). The key media that are considered in the story creation, often referred to as serious storytelling in the marketing industry, for brands and products include the mass media, computer games, computer graphics, smart media, simulations and virtual training (Lugmayr et al., 2017). Studies in marketing show that narrative advertisements are more effective than factual advertisements (van Laer et al., 2019). Of course, the traditional media of storytelling, novels and fiction from the book publishing companies, including Bertelsmann Stiftung of Gütersloh, Germany, are not driven by the commercial success of place promotion. The brand for them is far more granular, e.g. particular famous authors, and more heterodox, e.g. their Penguin Books brand in the United Kingdom and the United States. Certain book publishers have been acquired by larger mass media companies, for example, the publishers of Hilary Mantel's UK bestsellers are Fourth Estate, owned by News Corp, New York (Figure 3.2).

According to Lugmayr et al. (2017), serious storytelling from a media studies viewpoint can be visualised using their 4C mode. The four Cs represent the essential components of: Context, which refers to situation; course, which is the sum of plot; content, the actors and setting of the narrative; and finally channel, which is the delivery technology for the exchange of stories, e.g. YouTube, Facebook, a printed novel, Instagram, tweets on Twitter.

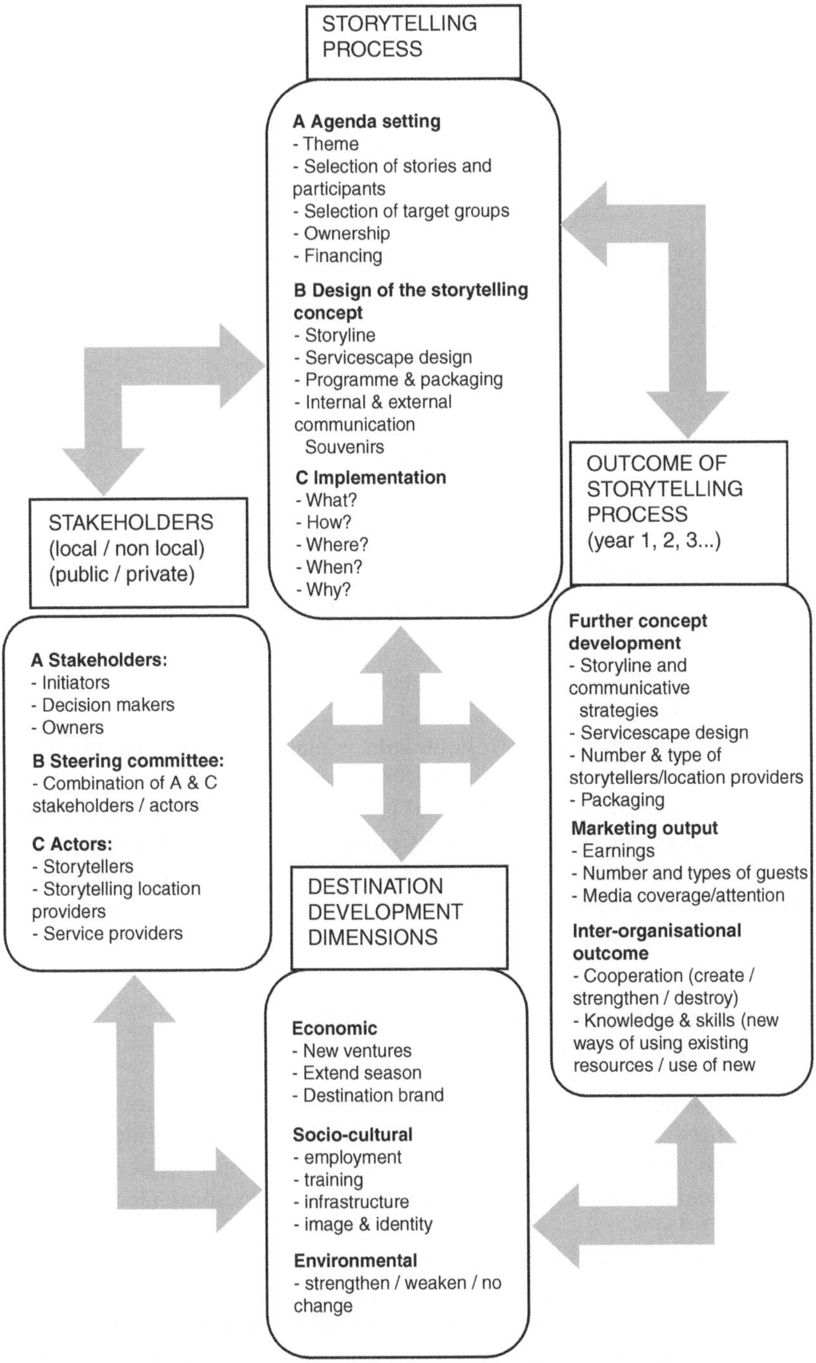

**STORYTELLING PROCESS**

**A Agenda setting**
- Theme
- Selection of stories and participants
- Selection of target groups
- Ownership
- Financing

**B Design of the storytelling concept**
- Storyline
- Servicescape design
- Programme & packaging
- Internal & external communication
  Souvenirs

**C Implementation**
- What?
- How?
- Where?
- When?
- Why?

**OUTCOME OF STORYTELLING PROCESS**
(year 1, 2, 3...)

**STAKEHOLDERS**
(local / non local)
(public / private)

**A Stakeholders:**
- Initiators
- Decision makers
- Owners

**B Steering committee:**
- Combination of A & C stakeholders / actors

**C Actors:**
- Storytellers
- Storytelling location providers
- Service providers

**Further concept development**
- Storyline and communicative
  strategies
- Servicescape design
- Number & type of storytellers/location providers
- Packaging

**Marketing output**
- Earnings
- Number and types of guests
- Media coverage/attention

**Inter-organisational outcome**
- Cooperation (create / strengthen / destroy)
- Knowledge & skills (new ways of using existing resources / use of new

**DESTINATION DEVELOPMENT DIMENSIONS**

**Economic**
- New ventures
- Extend season
- Destination brand

**Socio-cultural**
- employment
- training
- infrastructure
- image & identity

**Environmental**
- strengthen / weaken / no change

**FIGURE 3.1** Storytelling process. Adapted by Mansfield & Potočnik Topler after Mossberg et al. (2010).

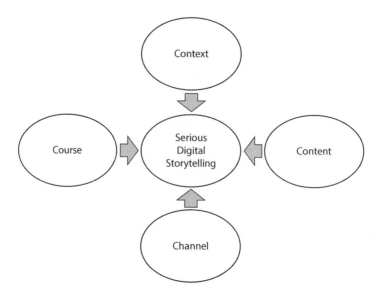

**FIGURE 3.2** Serious digital storytelling – the 4C model. Adapted by Mansfield & Potočnik Topler after Lugmayr et al. (2017).

In the digital era, marketers increasingly employ storytelling techniques to persuade their customers (van Laer et al., 2019). The application of storytelling in Tourism via smartphone apps is playing an increasingly important role, since stories told through social media are available in real time, dynamic, visible and, thus, influential (van Laer et al., 2019). Mobile storytelling, for example, offers many possibilities, including 'sending customers personalised information' (van Laer et al., 2019, 137) about the backstory of some cultural or tourist attraction. When it comes to education (in Tourism), storytelling can provide relevant learning experiences (Lugmayr et al., 2017).

### The relationship between journalistic writing and destination promotion

Tourism and media are deeply intertwined through co-creation (Månsson, 2015); Månsson (2015) elaborates on the mediatisation of tourism by observing that tourists are no longer only consumers, but producers of media content, creators and active participants of place branding (Jenkins, 2006; Månsson, 2015).

Mee (2009) argues that travel writing and journalism have a long-existing connection and emphasises that texts that are more journalistic in nature are often characterised as inferior to more literary ones. However, it must be observed that travel writing is not the same as travel journalism, because journalism must stay factual (Tiede, 2016). The so-called golden rules of journalism, which are considered by journalists at all times, are the following: Reporting must be accurate and obey the six news values

(1. timeliness – recent events are more newsworthy than less recent ones; 2. proximity – stories from the home region have a higher news value than those from distant places; 3. prominence – celebrities and well-known people have a higher news value than ordinary people; 4. uniqueness or oddity – bizarre and strange stories are more newsworthy than ordinary stories; 5. impact – stories that impact a large number of people have a higher news value than those that impact less people; 6. conflict – stories with conflicts and violence are more newsworthy); the story should start with the most important facts and end with the least important ones (the structure of the inverted pyramid) and answer the five Ws (Who? What? When? Where? Why?) (Writing-center, 2020).

In tourism 'as a major global cultural industry' (Heller et al., 2014, 432), the relationship between journalistic travel writing and destination promotion by the DMO is crucial. Professionals in tourism are aware of the role of contemporary media and communication concepts that are significant for the positioning of destinations (Potočnik Topler, 2017). When it comes to destination branding and promotion, the new media have an outstanding role (Potočnik Topler, 2019); in fact, they are considered as one of the most important trends that have marked the tourism industry and brought significant changes in the tourism communication (Leung et al., 2013, 3). As early as 2009, Mangold and Faulds (2009) wrote about consumers turning away from traditional media. Bruhn et al. (2012, 770) claim that social networks or microblogs are 'increasingly replacing traditional media', and that social media are an important source of information about different brands. Brand communication, and, consequently, destination communication, is not generated only by companies and DMOs, but growingly 'by the consumers themselves, through so-called user-generated social media communication' (Bruhn et al., 2012, 771), where consumers are also content creators. What is more, social media have changed planning and consuming travelling dramatically (Hudson & Thal, 2013; Zeng & Gerritsen, 2014), and it covers the whole process of travelling – from choosing a destination and servicers to sharing experiences (Zeng & Gerritsen, 2014). In the last decade, many researches have dealt with the role of social media in Tourism (Wilson et al., 2012; Leung & Bai, 2013; Stepchenkova & Zhan, 2013). However, when searching for holiday ideas and a new destination, buyers of tourism product will start with web searches and be served the sources from channels where investment in marketing campaigns has been heaviest. These channels are still dominated by the travel sections in the national press of the country of the IP address, language and currency of the searcher; this is referred to as eCommerce localisation.

A key stakeholder who is concerned with the perceptions that the searching public have of a city is its city council. In larger cities, like Nantes, which is examined in detail below, this responsibility for the destination image of the city is devolved to a specialist group within the council. In the United Kingdom, this department is given the abbreviation of DMO, the destination management organisation. The role of smaller DMOs is to promote the city, that is, if increased visitor numbers from a particular demographic segment are deemed important by the elected council. Larger, better funded DMOs also make infrastructure changes to encourage or discourage certain demographics. Often, too, this role has links with the encouragement of inward investment into a region, which leads to property development. The DMO, along with businesses who trade in the visitor economy, for example, hotels, tour companies, restaurants, museums and visitor attractions, try to understand and then to influence the destination image that tourists hold of their town. Nantes is chosen for this illustration of monitoring the effectiveness of place branding. The UK perspective is taken because although tourists from the United Kingdom know the Loire Valley as a destination the city of Nantes, it is not that well known, despite it being the sixth largest city in France.

As explained above, the travel section of the UK national press is the main channel to communicate the brand of Nantes to attract British tourists for short breaks. With English-language newspapers online for readers to browse, DMOs have an accessible source of data to evaluate how their city brand is presented. Of course, the DMO has also worked hard to provide opportunities for journalists and freelance travel journalists to write that collection of articles that can stay available for around 5 years, unlike print newspapers and printed colour supplements. In this illustrative study, a collection of travel articles is built as a database and then analysed into themes for comparison with the elements employed in place branding. The analysis discovers that three key elements used by commercial and academic place branders are not covered by the contributors to the travel sections of the UK national press.

### Collecting travel journalism for monitoring the brander's message

As a case study, a collection of 12 travel articles was gathered from online sources by simulating the searching and browsing by an English-speaking reader with a UK IP address and a currency of GBP, considering a short city break. The focus is on the French city of Nantes, on the river Loire, as a likely destination for UK visitors flying from Bristol or from Southampton. The content of the articles is analysed thematically for comparison with the elements used to give place value in city branding (Zakarevičius & Lionikaitė, 2013). The aim is to discover if the practice of travel journalism presents the same elements in the places they describe that city branders want to present to convey the image of their detination. If not, then which elements from place branding are not covered by the journalists and freelancer travel writers for the travel sections of the UK national press? Recommendations are then made from the analysis to the main stakeholders involved in this communication process, including travel section editors and their travel writers, and destination managers, their branding analysts and their press officers.

Place branding does not start with tag lines like 'I AMsterdam' or the original I♥NY, first written in French by Sartre in his travel book *Situations*

*III* (1945, 2). The analysis phase of place branding is well-documented in the academic literature (Azevedo, 2009; Zakarevičius & Lionikaitė, 2013). Azevedo presents a plan for asking residents why their city has a good quality of life. This approach of eliciting the emotional value to be found in an urban space finds a parallel in the aims of promotional travel journalism. As the analysis stage of place branding is examined, it appears to provide ways of working with local knowledge that is also used by freelancers for researching travel articles for the mass media (McGaurr, 2012). This differs from the social media company, TripAdvisor, which solicits consumer reports to encourage electronic word-of-mouth recommendations or criticisms. Urban heritage tourism has already been shown to provide value or satisfaction for the visitor where strong place identity is coupled with continuity (Ginting & Wahid, 2015). Continuity, as a place-branders' concept resides in the heritage architecture of historic buildings and to some extent in the memories tourists have from spending time in clearly defined urban areas, for example, from dining in a particular square (Ginting & Wahid, 2015). The research phase of travel writing is also able to discover and make explicit the history of this heritage architecture since it is considered part of the tourism product and hence made visible through promotional documents and web-pages (Benur & Bramwell, 2015; McKercher, 2016). The recovery of memories of time spent in urban space provides a greater methodological challenge, however, and may be too time-consuming for freelance writers for newspaper travel pages who often only have funding for two nights on-site (Saunders & Moles, 2016). Thus, the place brand that has been carefully researched and constructed by the destination management organisation may only be partially communicated to its potential visitors. Indeed, it is mainly what is called tourism product, or ground product, that is covered in the newspaper articles of the travel section, along with means of transport taken from the country of departure. In passing, in the UK press, this is often provided free to the travel journalist and declared at the end of the travel article.

Let's pause for a moment to look at the opening or incipit of one of the travel articles, to see what is being communicated and to imagine the writer, and those stakeholder pressures surrounding her. Alas, the original source of this article has now been removed from the web-pages where it was stored, since its site has had a refresh.

Nantes: A Dynamic City – Pays de La Loire – Brimming with Holiday Ideas

For those with a gastronomic voyage in mind, there are a multitude of choices of cuisine from traditional French 'plats' to contemporary fusion. The 'Bouffay' quarter is bursting with restaurants and a walk down its streets allows you to savour the aromas before making your choice.

If you are thinking of trying to replicate some of the delicious meals you have tasted, there are several markets where you can buy all the ingredients you need. Check-out the Talensac markets every weekend and we guarantee you will be spoilt for choice.

So you've got water, history, beautiful buildings, amazing exhibits and tantalised taste buds, but what about … shopping? You will not be disappointed! Those in the know 'Crebillon' on the weekend. The Crebillon street is so well known for shopping that its name has become a verb (Crebilloner) for one of our favourite pastimes, retail therapy! Here you will find an abundance of French and international brand names.

The headline title aims to situate the city of Nantes in its relationship with the Loire Valley. The Loire Valley is a space which tourists from the UK know of, but often cannot place on the map. The French name, *Pays de La Loire*, inserted into the headline title is the name of the region; to most UK readers this is meaningless, but it reveals that the author or the editorial team were French and wanted to show the DMO of the regional administration that they had been name-checked. District names within the city, the Bouffay quarter, Talensac and Crebillon, tell the readers at the city's DMO that Nantes has been well-mapped to encourage visitor footfall and commercial consumption. This is the role of this discourse practice; it is part of the cycle of economic production, promotion and consumption. The writer is making a living, probably within a publicity-writing company, a content marketing company or a local government organisation. It is a valuable text for visitors because it takes them rapidly to the urban spaces ready to receive them and make their short stay enjoyable. In passing, too, the writing gives the sound of insider information, with the phrase 'Those in the know 'Crebillon''. This adds value to the tourists' holidaymaking, by enriching their experience with local knowledge; the demographic of British visitors who find their way south to Nantes are interested in the French language and French lifestyles, as an exotic culture to enjoy and as a mark of social status.

### Tourism research that examines the role of travel writing in tourism

Researchers are attempting to discover where travel writing or literary writing has contributed to the increase in value of a place for visitors. Richards and Wilson (2004) propose that travel writers, and literary authors who write about their travel, for example, Kerouac (1922–1969), do influence visitors. This raises the question of whether the tourist wants to retrace the journey or to emulate the life of the famous writer when they go on holiday. Writing offers a way of negotiating identity when new cultures are encountered by the visitor, whether the visitors are reading a piece of writing set in their holiday destination or writing a journal or blog for themselves. Published travel literature does change tourists' behaviours, for example, Macaulay's (1949) *Fabled Shore: From the Pyrenees to Portugal*, is an example of literary travel writing becoming accidental promotional writing (Dann, 1999). The above analysis, although it uses the term travel writing to categorise the practices it examines, is more concerned with travel literature. It may be that the strong narratives contained within travel literature are a key feature that contribute to their use by tourists. The category of newspaper travel journalism is not analysed by the above research, leaving a gap in the knowledge on how the content of travel journalism can influence visitors. Clearly, the travel journalists who are working on the creation of these articles have in mind the behaviour of tourists and how their texts might change that behaviour. For example, in the piece we examined earlier, the travel writer begins one sentence in the imperative form of the verb, 'Check-out the Talensac markets'. This instruction or command reveals her identity to us, her job, as she sees it, is to exhort her reader-visitors to modify their behaviour in the tourist space of Nantes.

How these types of media-based travel articles are generated and their impact on destination image has been examined at a broader level (Gladwell & Wolff, 1989; Dore & Crouch, 2003),

> [These PR activities hosted by DMOs] may be known by a variety of names including press tours, visiting journalist programmes (VJPs) and travel writer or media familiarisation tours. They are all designed, however, to attract a variety of journalists, writers, film crews, documentary producers, etc, to a location to experience the tourism product and provide media exposure for a particular destination.
> *(Dore & Crouch, 2003, 140)*

Little is available linking the more contemporary concept of place branding with travel articles from the national press, except for the work of Kubacki and Skinner (2006) which explores whole country branding through UK media travel articles. They put forward their justification for using 40 articles from UK broadsheets to examine brand using these two positions to justify their research and findings:

> national brand is communicated through a nation's geography, history, proclamations, entertainment industry and media, art and music, its famous citizens and other features, as well as through its culture, language, tourism and travel writing and its commercial branded products.
> *(Kubacki & Skinner, 2006, 288)*

> Travel writing is a relevant data source as it tends 'to be more readily believed and accepted when the channel of communication used is itself a cultural artefact, rather than paid for media' citation from Anholt (2002).

To conclude this section, it can be seen that researchers do believe there is a causal link between travel writing, both in the form of travel articles from the media as well as travel literature from iconic authors and the behaviour of tourists. Travel writers and their employing organisations seek to modify this behaviour, either to encourage them to make destination choices in the first place or as visitors at the destination to make decisions on what to see and do in the city space. Value, pleasure and satisfaction are only mentioned in passing in the academic literature, the overarching consideration is motivation and place promotion to increase visitor numbers. However, a piece of literary travel writing is considered to have the status of a cultural artefact in itself (Anholt, 2002), and for that reason, it can be assigned a value. Thus, the writing or the author provides leadership for the visitor on where to go and what to do in the urban space. Because of these two points of influence, the delivery of the text needs to be made to coincide with when the visitors make their decision: destination choice when at home, activity choice when on holiday.

### The analysis stage during tourism destination branding

At the outset of a destination branding process, local knowledge is elicited from internal stakeholders to understand which aspects of a resort or urban space give its citizens a feeling of value, satisfaction or pleasure. Participation in this destination image creation, including the integration of local residents' voices is well-documented (Eshuis et al., 2014; Zenker & Erfgen, 2014). The first authors identify three key contributions from citizens in this brand development process: 'emotions, feelings and stylistic preferences' (Eshuis et al., 2014, 156). Zakarevičius and Lionikaitė (2013) go further in their examination of place branding to provide a list of elements, in order of importance, that are regularly examined by place-image researchers in questionnaires or through consultation with citizens. Indeed, this synthesis by Zakarevičius and Lionikaitė (2013) is particularly relevant to travel writers and DMOs since it considers the work of both academics and practitioners in the field of city branding, ranking the findings based on analysis of both these groups. From their research, reproduced below, a list of themes can be proposed for the coding of the writer's content in the travel articles on Nantes in this study (Table 3.1):

They argue that campaigns that do not include the interests of local people lead to a loss of authenticity and a lack of place recognition in the resulting promoted image, when the brand is subsequently communicated (Zakarevičius & Lionikaitė, 2013). This communication of the place brand includes newspaper travel articles. The promotion of what is already there in a place, its geography and weather, is almost directly opposed to writing advertising copy, where the buyer's value system is the main consideration, since in advertising, it is modifying the buyer's behaviours that is at stake in achieving the sale of a tourism package.

Azevedo (2009) conducts city brand research using citizen involvement. Azevedo's questionnaire, administered to 107 subjects (Azevedo, 2009, 8), elicited the positive elements of the town and reports back these elements in the language used by his subjects, for example, 'welcoming' (13.3%), 'beautiful' (11.1%). A relatively high 11.3% of the subjects mention the mayor as contributing positively to their feeling for the town, thus confirming points

**TABLE 3.1** Place-branding analysis coding scheme

| *Numbered brand theme* |
| --- |
| 01-heritage |
| 02-leadership |
| 03-infrastructure |
| 04-culture |
| 05-people |
| 06-society |
| 07-government |
| 08-events |
| 09-history |
| 10-perceived-value |
| 11-social-conditions |
| 12-public-services |
| 13-stories |
| 14-business |
| 15-environment |

*Developed after Zakarevičius and Lionikaitė (2013, 152–153).*

of leadership and government from the list of elements reproduced above (Zakarevičius & Lionikaitė, 2013). Very useful, too, to the travel writer and place brander alike, are the identifiable locations that locals associate with these positive feelings, e.g. 'excellent beaches', the 'wine cellars' and the 'river Douro banks', since the local subjects are more likely to be drawing on their positive personal experiences rather than remembering advertising copy from promotional literature (Azevedo, 2009). The researching DMO and the travel writer could then go to investigate the physical locations in the town that have these dramatically positive associations for people.

Azevedo (2009) does not ask his subjects for stories, even though he asks about heritage, yet Zakarevičius and Lionikaitė (2013) rate stories, in a combined grouping with heritage, as highly important. Finding ways of eliciting stories from locals linked to their positive locations would, therefore, be a worthwhile element to investigate. The primary skill of destination-image researchers lies is their ability to design discovery instruments which are powerful and yet sensitive enough to record the positive components of identity formation of people who have found aspects of value and culture in the urban space under study. An intellectual movement did exist for this discovery work, the Situationists in the 1950s.

### Existing destination branding for Nantes

In Nantes, a public company was formed in 2011 to manage a tourism and cultural promotion project. This DMO company is called Le Voyage à Nantes, LVAN. LVAN has an annual budget of 27,000,000 euros and 260 full-time equivalent employees (Nantes Tourisme, 2015). Le Voyage à Nantes is une société publique locale, an SPL, a legal entity designed to make the management of public funds more competitive in Western democracies. Under an SPL, the managers and employees, at least in Le Voyage à Nantes, consider themselves free to commission artists and to act as sponsors, rather like Renaissance patrons of the arts. This assumed role is interesting for the professional travel writer and this project since it offers a way of financing guided walks or commissioning the research, writing, packaging and distribution of pieces of travel literature or even of workshops or residencies around place-making in Nantes.

To give context to the tourism activity of UK holidaymakers in Nantes, data from INSEE's Pack Hôtels Product (INSEE, 2016) were used, first to examine the number of nights stayed in the whole of France by tourists whose country of residence was the United Kingdom. Examining the figures in detail, it can be seen that the fall in numbers of UK residents using hotels in France between 2008 and 2011 is probably due to the world economic depression after the banking failures, which became known during the autumn of 2008. From its low point of 10,468,822 nights stayed in hotels in 2011, as the bar graph above shows, some recovery can be seen up to the most recent figures available (INSEE, 2016). Statistics for hotel night-stays for Nantes were available from the DMO, Le Voyage à Nantes, and although not covering the same complete period, do show the interest that UK tourists have for Nantes. Between 20,000 and 30,000 nights were spent per year in Nantes by tourists from the United Kingdom. Over a five year spread, this fluctuation of plus or minus 10,000 nights means that different conditions can affect the numbers dramatically. Even if each night represents a spend of only £80, that is still £800,000 variation according to changes, which may include promotional writing and publishing.

The urban area covered by the data is the CU, the Communauté urbaine, which comprises Nantes plus the 23 suburban spaces around the city. An increase in visitor-nights spent in 2013 may be attributed to the award of European Green City 2013 to Nantes and the additional traffic that would be created, particularly from Bristol. Bristol was bidding for the next Green City award and planning a new air route from Bristol to Nantes. If those two batches of data are combined, it does indeed show that the Green City year of 2013 gave Nantes an increase in the total share of UK visitors, against all French regions.

This percentage share measurement is also a reminder that French DMOs compete with other towns across France for their share of inbound British visitors; this demonstrates the importance attached to city branding and the measurement of its effectiveness. Thus, analytical techniques such as the one explained in this book will be valuable tools for the DMO to gauge the effect of their branding and identify gaps in the place brand image.

### Monitoring place branding for the DMO

In a commercial environment, travel journalism may be used to increase revenue by promoting the tourism product available for those on a city break, but for the visitors themselves, cultural knowledge gained from reading and exploring heritage is linked to personal emancipation. Cultural tourists seek out experiences that will free them to enjoy a more fulfilled life. Personal value for holidaymakers may derive from accomplishments and positive identity growth; these are related to uncovering the hidden experiences in a modern city (Koeck & Warnaby, 2015).

Psychoanalytical practices also contribute to the value that would-be visitors seek; Bernstein explains that if people recognise their own self in a proffered narrative, they can continue their interrupted self-formative process and gain emancipation (Bernstein, 1995, 64). In simpler terms, this is close to identifying with a narrator in a story, whose actions then influence the readers.

## Data integrity processes for the travel journalism articles

The method of data collection for this brand-watch can be easily repeated by DMOs and travel writers. The articles are captured from the Internet. Two types of simple search are used to simulate an English-speaking, UK-based user looking for a cultural holiday or for city break information on Nantes. The search engines, Google in Chrome and Bing in Internet Explorer are used. Only the text of the articles is captured, then cleaned by saving as UTF-8 encoded format to preserve any accented characters. Any article clearly dated older than 2000 is discarded. If no date can be determined, then date of capture is applied instead.

Keywords for Searches: Nantes, Nantes Travel, Nantes Culture, Nantes Travel Writing, Nantes Travel Article, Nantes City Break.

## Methods for coding the travel articles for analysis

The framework method is often used for discovering tourism knowledge (Brunt, 1997), and Gale et al. (2013) provide a guide to using the framework method for thematic content analysis of qualitative data. Their process takes five steps from coding through charting onto a matrix. Their final step uses interpretation through memo-writing. These steps employ inductive reasoning, where the data suggest strong evidence. The final step uses abduction to move from observation to plausible new theory from the findings. Gale et al. (2013) use open-coding since they have no specific pre-formed hypothesis about what they will find. However, the brand-watch analysis concentrates on finding matches to the coding list from place-branding practice to answer the question: Do these easily-found travel articles communicate the DMO's message? And, where are the gaps? That is, which place-branding elements are poorly represented in the publicly accessible travel articles on Nantes?

As a match is found in each travel article, the researcher adds it in the correct cell in the spreadsheet and then changes the tally to a total figure on completion, e.g. IIIII = 5. Food is coded in 04-culture and terroir in 15-environment. With a sample of 12 articles, the coding and adding the tally to the spreadsheet can be completed in a day's work. Here, Table 3.2 provides detail from the spreadsheet of these travel articles.

TABLE 3.2 Extract of spreadsheet for recording branding elements found in each newspaper travel article

| Number-coded article title | A: 01-heritage | B: 02-leadership |
| --- | --- | --- |
| 01 KLM Strolling in Quartier Graslin | 1 | 0 |
| 02 Arrival Guides Welcome to Nantes | 1 | 1 |

Explored visually, it can be seen where the DMO's message is finding its way through the main online newspapers and travel media to readers in the United Kingdom. The zero data or empty columns show where no communication is happening.

The richest article, in terms of matching the place-branding elements, is by Audrey Gillan and was published on 13 August 2015 online in *National Geographic* and updated on 5 July 2021. It is in a magazine, of course, rather than a newspaper travel section, and *National Geographic* do have a reputation for well-researched writing. Gillan starts by using the form of a diary entry or blog post placing her readers there with her at 8 AM in the market. Notice how she introduces locals by name as characters in her article. Then how she links the local wine with the geology and geography, this is summed up by the word terroir. Gillan name-checks a local restaurant and gives her readers a wine type that will help them when ordering in Nantes, Muscadet.

> It's 8am in Nantes' Talensac market [...]
> Ebullient and charming, Bourget, owner of La Raffinerie in the Madeleines area of this north-western city, shows me fat radishes at a stall where the sweet smell of Gariguette strawberries prompts a kind of Pavlovian pulling of purses from pockets.
> [...]
> Meanwhile, at Vins de Loire, Solène Franquet teaches us that muscadet is a wine of far greater variety than we'd rather ignorantly assumed. We taste an example that's vivid and fresh, with an air of the sea, and another that's more viscose, a 'third level' crus communaux distinguished by its candied fruits and honey.
> 'There are 600 wine growers of muscadet, most of them in Sèvre et Maine, but the soil is very different so you can have very different tastes,' says Franquet. 'A young muscadet makes little pearls on the tongue, while a crus communaux has a very special soil and is aged for 18 months.'
>
> *(Gillan, 2021)*

### How do the travel articles match the 15 place-branding elements?

The analysis reveals the relationship between the place-branding elements and the themes covered in the travel articles from online newspapers. The matrix graph shows that one travel article addresses 5 of the 15 elements from place branding, a very high correlation. The graph does reveal some consensus by the travel journalists, for example, 04-culture and 09-history are elements that are each covered in 7 out of the 12 articles found on Nantes. Whatever conditions affect the writers' choices in what to include in their writing, a visible pattern does emerge. Close-reading of the references to 04-culture shows lists of where to eat, which are very similar to those found in guide-books. The limited time the writers can spend in the field and their perceived expectations of their audience, both their editors and their readers determine what they include from the tourism product. No attempt is evident to provide a story for readers to incorporate into their own narrative in any of the pieces, although a suggestion of a narrative is made by Coates when he introduces the character of his 21-year-old son into the article using dialogue (Coates, 2014) and a certain resolution occurs to the story in the carousel scene (Coates, 2014). It must be concluded though that online travel articles published by high profile media companies do not provide opportunities for readers to identify with Nantes or to take ownership of the city through narrative or stories.

Notice that three branding elements are not covered by any of the articles, these are 06-society, 11-social-conditions and 13-stories. A documented example exists of how these three missing elements can be researched and found; the academic sociologist, Delcroix uses reflexive forms of action research and recounts the story of one of her projects in Nantes (Delcroix & Inowlocki, 2008); essentially, it is a success story where isolated fathers in the community are empowered to engage with teenage groups to jointly develop social capital to improve the social conditions of the city. Through life-story narrative sessions, the fathers create family memory and from this gain power through increasing the networked social capital of the urban space (Bourdieu & Passeron, 1990). The methodology of narrating life-stories using practitioner-led reflexive sessions is explained more thoroughly by Chaxel et al. (2014), their study shows how the case participants can reveal resources from earlier times in their lives that can be used as part of emancipatory practices in the future (Chaxel et al., 2014). Delcroix and these other academic researchers do spend far longer in the field than travel journalists can spend when on assignment for online media publishers, though.

### How can these findings inform the process of commissioning travel writers?

Commissioned travel writers and the field researchers in place branding lack direction to the urban places in which element 13-stories are being made, for example those complex ones from social conditions discussed above (Delcroix & Inowlocki, 2008). The tourists' places, too, elicited from the hidden delights of a modern city (Koeck & Warnaby, 2015) could also be added to the places for commissioned writers. It opens the question of how the places are distributed across the space of the city under study, Nantes. Michel, for example, explores the creative quartier of Nantes, called Les Olivettes in his doctoral research using semi-structured interviews and from that delineates a plateau of interest in the city by mapping the networks of relationships between creative companies (Michel, 2014). Saunders and Moles (2016) present an operationalised process for eliciting stories of place in their data collection of audio recordings by young men in a district of Cardiff. These long-term academic research approaches could be developed into a more compact process for engaging with locals focussed on specific locations in the city. In this way, stories of value to others who want to enjoy the places could be mobilised as part of a more accessible network of social capital.

### Recommendations to destination-image researchers, commissioning editors and travel writers

The DMOs for towns and resort areas could continue this ongoing analysis of the travel articles that are published online for UK readers using this model. This monitoring process identifies which elements remain ignored in the travel press. By commissioning writers, the DMO could specify which elements to include to engage the audience with the city. Identifying a leading writer in this field requires a knowledge of the media scene in the country of the outbound tourist group. The element, 13-stories, in place branding, can be found in the cultural history of the town, which coincides with branding element 06-society, and these are present in the architecture and literary texts set in the town.

# 4
# VALUE, AXIOLOGY AND WRITTEN REPRESENTATION OF PLACE

## Can place be shared using written texts?

Defining place challenges theorists. One of the reasons for that is that many disciplines share the concept of place, which is elusive in its nature, thus, perceptions and ideas of a place differ widely. As does terminology; 'place', 'space' and 'cultural geography' are all terms associated with places (Shi & Zhu, 2018, 224) and also with tourism destinations. The question of place is sociological, philosophical, political, economic, and emotional. In tourism research, for example, place attachment (Dwyer et al., 2019) and place construction are drawing the attention of researchers. Brown (2005, 18) discusses the concept of place as 'a social construction formed by human-environment interactions'. Besides the concept of place, authors also discuss the sense of place, which reflects many things and aspects from the cultural, religious, historical and personal to economical (Stuart Chapin & Knapp, 2015). Place is a room, even a couch, a bed, a writing desk, a house, a village, a town, a city, a bridge, a clearing in the middle of the woods, a special rock on the beach, an exact apple tree in the meadows, a garden or a park. In research literature, several definitions of place are found: from 'the physical aspect of the environment at hand' (Jeremiah, 2000, 23) to 'the setting in which issues of writing and other language-related skills are housed and discussed' (Jeremiah, 2000, 23). Patterson and Williams (2005, 361) write that the concept of place emerged in the late 1960s 'as a prominent focus for exploring the relationship between humans and the environment', and point out that there is still a 'lack of conceptual clarity'. Place has its own physical, historic, natural, social and economic characteristics, and, as such, it forms its own relations with people, organisations and other places. Van Assche et al. (2020) argue that every place also has its own narratives, while Grenni et al. (2020) point out the importance of cultural narratives in place branding,

DOI: 10.4324/9781003178781-5

which 'refers to the creation of value in space by reinforcing and representing place assets in a cohesive manner, as a narrative image of the place itself' (Grenni et al., 2020, 1355). Stories are deeply entwined with places, also because every story has a place of happening, and the identity and image of every place are also constructed by stories (Mundell, 2018). On the other hand, places can be created – skilful travel writers and branding experts can create a place which is significant from the perspective of Tourism studies. Another significant aspect of places is that places have different meanings (Gustafson, 2001) and that people develop attachment towards them. Usually, individuals are attached to places where they grew up or where they spent good times. According to Stedman et al. (2004, 582) experience counts, and they claim that place attachment 'is built through experience'. Hidalgo and Hernandez (2001, 274) discuss the concept of place attachment and state that the 'terminological and conceptual confusion' of the research area hindered advances. Patterson and Williams (2005, 375) explain that many dimensions of place concepts exist and that some of these dimensions can be 'operationalised quantitatively', while others 'are explored more subtly through qualitative approaches'.

Smith (2015), Ram et al. (2016) and D'Orey et al. (2019) studied places from the tourism perspective as the central tourism category, essential in tourism product development. More precisely, Smith (2015, 220) explored 'the importance of place in the context of tourism product development and marketing' and established similarities with *terroir*, and he named it 'the *terroir* of a place' that includes 'history, local traditions and cultures, religion, industry, the natural environment, cuisine and arts, as well as attractions and events' (Smith, 2015, 220). Smith (2015, 220) argues that place is a complex category, a lot more than a mere geographical category or term, and that a 'key feature of place-based product development and promotion is the identification and telling the story of a place through a variety of narrative techniques', such as storytelling, print, video, graphic, digital media. Ram et al. (2016) focused on attractions as factors of place attachment and authenticity and established that tourism attractions contribute to place attachment and authenticity. D'Orey et al. (2019) believe that understanding the relationship between places and tourists is vital in Tourism, and it leads to improving the tourism sustainability. D'Orey et al. (2019) also established the following four major concepts concerning the perception of 'sense of place': (1) Belonging to Place (Relph, 1976; Chang, 1997; Kaltenborn, 1998; Vorkinn & Riese, 2001), (2) Affection for a Place (R. C. Stedman, 2003; R. Stedman et al., 2004; Yuksel et al., 2010; George & George, 2012), (3) Commitment to a Place (Shamai, 1991; Smaldone et al., 2005; Shamai & Ilatov, 2015) and (4) Identification with a Place (Proshansky, 1978; Proshansky et al., 1983).

The concept of place is, of course, related closely to the concept of time (Friedland & Boden, 1994), and the two concepts together represent the framework for The German expression 'Zeitgeist', which, nowadays, is used to describe the spirit of certain times in certain places, while according to Gieryn, 'place persists as a constituent element of social life and historical change' (Gieryn, 2000, 463). In tourism development, places are sought out that have the attributes suited to become attractive to visitors because they have 'the unique features of a place that reflect its history, lifestyle, or environment' (McKercher et al., 2004). Więckowsky (2014) argues that tourism space differs from other spaces due to different activities which define space; thus, inward investment, human practices and communication have a significant role in development of a tourism space. Tourism space is, in fact, 'part of the geographical and socioeconomic spaces where tourism phenomena occur' (Warszyńska & Jackowski, 1979, 31 in Więckowsky, 2014, 18), and it is 'a subspace of general geographical space, i.e. made up of natural and social components' (Liszewski & Bachvarov, 1998 in Więckowsky, 2014, 18). Very often, formal institutions, such as municipalities and management agencies, play an important role in the creation of places or, as Stedman et al. (2004, 583) put it, in 'the creation of place meanings'.

Van Assche et al. (2020, 1276) believe that 'realising fruitful synergies between spatial planning and place branding requires a thorough understanding of planning and branding practices, and the embedding of both in governance', thus confirming that elected local representatives are best placed to manage place-branding analysis projects in coordination with the planning application process for their cities and towns.

Attributes at the core essence of place, which are related closely to human perceptions, are dependent both on experience and observation (Brown, 2005) and were discussed by Ekblaw as early as 1937. Later, Tuan (1974, 1977), Relph (1976) and Brown (2005) tried to describe the cognitive, emotional and behavioural components of 'an individual's sense of place' (Brown, 2005, 18). According to Brown (2005, 18), 'humans are active participants in the landscape – thinking, feeling, and acting – leading to the attribution of meaning and the valuing of specific landscapes and places'. This theorising provides a basis for the place-making writer to interrogate their own participation in activities and attempt to relay their thoughts and feelings to their readership as a creative activity. For the researching place-writer, this is a form of personal anthropology, by asking what am I being here? And of personal ethnography, by asking what do I believe I am making here?

## Writing as a place inquiry: What can be known about cities through text?

Space exists in literature and 'represents a reference frame for literature' (Škulj, 2004, 22). For its existence, literature needs space, since it represents the setting of story events. In literature, place is usually related to time and to form, 'the social setting or the social context of a literary work' (Jeremiah, 2000, 23). Knoop et al. (2016, 35) state that when 'communicating the perceived aesthetic appeal of a work of literature to others, we project our experience onto the space of available verbal concepts and select those we find most appropriate to capture the nature of our impression'.

Space in literature may be entirely fictional and created by the author. On the other hand, the author may use reporting techniques to describe places and their everyday routines realistically. In the case study on Montreal, below, this is examined in detail to show how the literary author is combining social relationships with the spatialisation of the city, so that they each illuminate one another.

In Michel de Certeau's essay 'Walking in the City' (1980), he argues that urban places are brought to life by people walking. Morris (2004, 676) states that this essay 'offers a persuasive theoretical framework for understanding the temporal and spatial operations of popular culture, especially in terms of audience practices'. Walking, which may be an ordinary experience for the locals, may become an extraordinary experience for visitors and tourists. This is termed the 'extraordinary everyday' by Morris (2004, 676).

On the one hand, literary texts can communicate the emotions and the detail of the city, and on the other hand, visitors and tourists can learn by walking the city and 'reading' it. When the walker, who is also the writer-researcher, writes their observations about the city, the city is, in a way, labelled and characterised, and a certain perception is created which influences the image of the city. The future explorers of the city may be under the influence of this image, and all the previously noted images when reading the texts by previous writer-researchers before walking and researching the city for themselves. Texts produced by the writers-explorers can, or may, affect the city's identity and trigger and encourage discussion on the city's identity and its future development.

## Literary writing as place construction, a study of Proulx's Montreal

Monique Proulx is a Québecoise novelist and playwright, born in Quebec City in 1952 but living and working in Montreal. Her early work in the 1980s quickly gave her recognition with a collection of short stories (1983) *Sans coeur et sans reproche, nouvelles,* winning her the Prix Littéraire Desjardins. The work in the case study on literary place-making here is her first novel of the new millennium and her third novel to date *Le Coeur est un muscle involontaire* (2002), available in English translation as *The Heart Is an Involuntary Muscle* (2003). Proulx makes it clear that the novel is set in the city of Montreal and in the new century by mentioning dates, such as 2001, and by using street names with the familiarity of a local, 'rue Sherbrooke, coin McGill' (Proulx, 2002, 193).

The approach to her novel and literary writing as a discursive practice, in the following discussion, follows the position of James Reid:

> [There is] a modern awareness that discursive paradigms construct, rather than represent, geographical and social space.
>
> *(Reid, 1993, 19)*

So, although Proulx chooses a known cityscape and an urban society in a particular phase of its development, particularly during changing work conditions, the discussion below will explore and show how her literary practice is her construction of place, and urban place in particular.

Proulx's (2002) *Le Cœur est un muscle involontaire* is selected for this study because it is a new century novel set in the identifiable urban space of francophone Montreal. In particular, it uses twenty-first-century work patterns and culture to mobilise its characters and place them in urban settings where they can be observed in relation to the social structures existing in this contemporary commercial capital. One can, thus, legitimately ask how the author uses her character types to create this urban space and to react to the distribution of power at work there. From an analysis of their movements across space, their occupation of spaces and the spatial practices apparent in Proulx's writing, a critique of the use of urban space can be developed that will shed light on the author's own textual creation of the city and establish conclusions about the fears, desires and understanding of the city in contemporary urban culture. This is to show that contemporary literary travel writing can also use these writing practices to communicate a vibrant, living urban space rather than a nostalgic view of a past era which is now inaccessible to the contemporary visitor. For example, the literary travel writer can represent contemporary working roles in their stories as Proulx does with her characters working in new media web design.

This analysis is concerned with the use of the writing practices, and perceived social practices, and the way they are used to delimit, to control and to reproduce space, for example, the use of written and legal documents by a social grouping to authorise the use of the space. This use of legal documents has a close relationship with tourism development, where planning consent is needed to build or to change use of a building or area of land in the city. In the operation of businesses, too, legal entities such as partnerships and limited liability companies are regulated through state taxation and local business rates.

The novel revolves around Florence who is a 25-year-old web developer in a very small new media company; at start of the novel the proprietor, Zéno, is the only other worker. Florence acts as the first-person narrator throughout the story, so it is through her point of view that the characters of her father, Pepa, her employer Zéno, Pierre Laliberté, a reclusive novelist, and Zéno's dog, Poqué constrain or open up the space in which she operates. The opening chapter sets up a quest for Florence in that her father, Pepa, dies without having connected with her as a daughter. This quest for a paternal figure is subverted as the narrative progresses. The quest is set in motion when Florence realises that she has not even heard her father's last words as he lay in hospital, even though she spends a great deal of time in a bedside vigil.

The reader is allowed access to three key family groupings, one of which provides Proulx with the opportunity to look back at what may be Canada of the 1980s. These three families are: Florence's mother, father and two brothers in the old century, Zéno's mother and estranged father and

Pierre Laliberté's sister and his partner living outside the city limits. The family groupings and their cohesion provide a strong contrast to Florence's single status which runs throughout the novel. For Proulx's sensibilities, a solitary woman occupying her own space in the city is worthy of the central focus in her novel. This could be read as referring to Quebec's declining white francophone population. Even though the French-speaking population of Montreal gained the ascendancy by the end of the Quiet Revolution (1960–1976), it is a social group that is failing to reproduce its social space in the new century, to use sociological terminology. Proulx further explores this theme through the description of Florence's mother and brothers. This 1970s family is shown as a failed rhizome; it cannot propagate or produce fruit. Her mother, aged 58, even though she has a new lover, Alonzo, is not going to have any more children. Her brother Bruno is divorced with no children. Florence muses to herself:

> Why are we sterile? Why haven't we sent out living feelers, rootlets or rhizomes?
>
> *(Proulx, 2003, 40)*

Why does she use the spatial image of an underground rhizome rather than the aerial seeds or pollen of a flower for the future of French Quebec? Is it a social group that is already buried? The legal document of the marriage certificate is absent from the novel; in the year that most of the action is set, none of the couples is married. As a written or legal document, the marriage certificate evolved to secure property, land and buildings as a space for reproduction of a social group. As far as Proulx's Montreal is concerned neither the generation involved in the Quiet Revolution, portrayed by Laliberté, nor the emerging 20-somethings of the twenty-first century use the legal tool of marriage to secure its future, nor its communally held space in the city. If Maurice Blanchot's proposal of how a writer uses the idea of character has validity in this instance, then Proulx gives us a clear picture of a static social space of Montreal in 2001. Blanchot proposes that the invention of character is only another device of the literary text with which the writer tries to restore their relationship with the space of the outside world (Blanchot, 1955).

The Florence character displays many of the traits of the mythical Orpheus which prefigures her turn to writing by the end of the novel. A playful confusion of the myth is that Orpheus is taught to play the lyre, for this metaphor read 'taught to write', by Apollo who may either be his father or simply a suitor for Thalia. Later, Orpheus in classic literature ventures underground and then breaks the only rule by looking back. Proulx has her main heroine replay that scene (Proulx, 2003, 45–47) which initiates an underground spatial theme in the novel.

Examples of the underground theme begin to multiply. The latest novel written by the Laliberté character is called *La Périclitouze*, a portmanteau word which suggests the verb, péricliter = to be on the decline. Play is made on this French word in the title of the novel which forms part of the quest of the heroine and hero, Zéno. Zéno gives a copy of the book of declining to Florence for her birthday (Proulx, 2003, 44–45). All of her Chapter 3, 'Beneath the Earth' (Proulx, 2002, 18–35), and how she portrays writers as descending underground into the subterranean wells of their words (Proulx, 2003, 19) moves writers, in the created imaginary space of her novel, into the underworld. The burial scene of Zéno's dog, Poqué (Proulx, 2002, 29), and the fact that Poqué's name is related to name of Florence's father, Pepa, and how Pepa is buried from the start of novel, work to close off the past. The author Proulx removes the two characters from Florence's life that make the heroine's life untenable, the silent or non-speaking oppressors: Pepa, her father and Poqué, her employer's dog. The deaths of these two characters occur very early in the story and this opens up two key spaces for Florence. Her parents' home is ostensibly easier for her to negotiate and Zéno's car is easier for her to ride in. This latter space is of key importance for Florence since Proulx has not allowed Florence to drive herself, being in the cars of Zéno and, later of Laliberté, permit Florence a mobility around the city and out of the city that she cannot enjoy independently of these two men.

Thus, the two deaths seem at first to signal a removal of the old guard and the previous generation, liberating the heroine so that she can use the previously untenable spaces to increase her scope and power. As her use of these spaces develops, though, they become contested by the introduction of further characters.

Proulx depicts a recluse novelist and contrasts him with the web design work that Florence does for Canadian novelists who seek fame. The New Media industry in Montreal enjoyed governmental and private support in the new century, in much the same way as the railways, which gave nineteenth century Montreal links to US ports and Anglophone markets. Through the early 2000s, the establishment of la Société des arts technologiques (SAT) and its move to larger premises in 2002 were a signal that digital art could be consumed respectably as traditional art. Proulx in the novel is clearly troubled by this attempt to shift computing work to sit under the rubric of art. The main tension throughout the whole novel is between the production of the computer-worker and the art production activity of novelists. The space of these new workers of the digital age is nomadic or at least without physical importance; the web designers cannot afford office premises and only when successful, can the literary or artistic writers find space to work.

In conclusion, literary travel writers can draw the lesson that the inclusion of a strong theme, in this case the uses the novelist makes of the subterranean, creates a space that can be enriched by a more nuanced portrayal of the social, economic and political upheavals taking place in the city under study.

## Place-making with travelogue and with literary travel writing

Travel writers contribute to the image of places socially, economically and historically depending on the type of text they are creating; and it has been well established that travel writing plays a significant role in destination branding. However, a distinction exists between guide-book writing, travelogue and literary travel writing. The differences are in the aims of the piece and can be seen in the style, grammar and point of view in the text itself, taking, as an example, the city space of Trieste, Italy. The Welsh historian and travel writer, Jan Morris (1926–2020), knew Trieste from 1945 when a member of the occupying forces of the Free Territory of Trieste. Morris visited again later in life and in 2002, Faber published the travel book from Morris' writing (Morris, 2002). The text is valuable to this study, since it shows shifts between travelogue and the quite different form, called here, literary travel writing. This early scene-setting uses the present tense, which is associated with guide-books and travelogue. In travelogue, the present tense is used to describe city spaces and monuments that never change, a type of eternal present tense, for example scene-setting in this way: the main square is surrounded by eighteenth-century buildings and offers a viewpoint out across the bay to a small white castle that overlooks the sea. All the verbs are in the present tense, and this guide-book way of describing the space suggests that the square has always had that layout and always will have. Jan Morris draws on this guide-book present tense, but a subtle difference places this travel writing in the style of travelogue only initially; quite abruptly, it starts to show a feature of literary travel writing with the inclusion of a first-person narrator:

> a train clanks somewhere; a small steamer belches smoke; a band plays in the distance and somebody whistles a snatch of Puccini – or is that me? The heavily pompous buildings that line the shore, spiked and pinnacled with symbolisms, seem to be deserted.
>
> *(Morris, 2002, ix)*

As can be seen, the framing for this evocative cityscape piece above is set within Morris' reminiscing back to the 1940s. The character of the narrator is introduced, 'is that me?' (Morris, 2002, ix), which is a marked feature of literary travel writing, but it remains tentative. Not until pages 13 and 14 does Morris make it clear that a live travel story is unfolding, with possibly two protagonists staying at the five star hotel on the city square called the Piazza Unità. The narrative, however, continues in the present tense, so that the current journey remains undated:

> and here we are ourselves sitting at a table outside the Caffè degli Specchi, the Café of the Mirrors, which has been comforting its customers with coffees, wines and toasted sandwiches since the days of the Emperors.
>
> *(Morris, 2002, 13)*

For literary travel stories, one might properly expect a recounting of the activities in past tenses, for example: we found a table and sat outside the Café of the Mirrors. Readers, then, would be able to pinpoint the experiences enjoyed by the narrator and their companion. Morris does demonstrate a strong interest in the past of Trieste to illuminate the urban space, using historical figures to bring character to the built heritage visible to the writer and to tourists. In this next passage, the background to the museum of modern art is brought to life, with even a reference to an older guide-book, Baedeker's. However, the resulting paragraph remains as travelogue, rather than sharing Morris' own emotions of the gallery; nor does it take readers inside to experience a specific painting as would literary travel writing:

> Revoltella died in 1869 and left his house to the city, stuffed with the works of art that testified to his culture and his wealth – 'handsomely fitted up', commented Baedeker's *Austro-Hungary* approvingly in 1905. It has been enlarged in recent years to incorporate Trieste's civic gallery of modern art.
>
> *(Morris, 2002, 30)*

Morris mentions that the waiter tells them that the coffee is local. This presents an opportunity for literary travel writers who are interrogating the urban space to research further into the ethnobotany of coffee in this city. The results would uncover the coffee roasting family of Illy, and opportunities to include stakeholders in the travel piece. The evocative nature of coffee drinking, too, would enable the researcher-writer to interrogate their own memory for times when coffee, or specifically Illy coffee had been experienced. The literary writer shares personal tastes to offer opportunities for the co-creation of visitor experiences. By providing focus on the fragmentary moment of sipping hot coffee, the writer shares an activity that the visitor could easily reproduce. Further, if it is invested by the writer with a link to Trieste, then this authenticates the experience too. Finally, a large, well-established company like Illy of Trieste would probably be happy to provide details of its efforts to safeguard the environment which might provide further research material for a travel piece.

In preparatory literary reading for fieldwork in Trieste, the researcher should also turn to the writings of Boris Pahor (1913–2022) alongside those of Jan Morris. In his short travel piece, 'Place Oberdan' (Pahor, 2018, 63–79), journaled as written on 20 September 2008, Pahor uses personal memory to interrogate a square, called, Piazza Guglielmo Oberdan. It is only a kilometre north-east of the square of Morris' piece, so a walking route can easily be planned to connect the two research plateaus. Pahor offers an example of changing point of view as the literary travel writer processes their own experiences, even overnight. He recounts stepping into the atrium of the New Bank of Ljubljana (Pahor, 2018, 76–77), inside a building in which he had not set foot since February 1944 (Pahor, 2018, 70–71). He recounts painful memories and emotion from that time; however, the morning of his synthesis of the travel piece, his thoughts have a new interpretation of the square, that he then renames the square of Madame Oberdank, with a final -*k*.

Here, he recalls that the allied troops of occupation in 1945 authorised the radio station on the square to start broadcasting in Pahor's own language of Slovene. The catalyst for these moments of recall was his entry into the bank, and the minute detail of noticing a colourful new wall plaque. For the writer-researcher, the lesson is that it is often valuable to step inside even non-tourist buildings during reconnaissance of a new urban space to seek out new experience.

## Eliciting knowledge with the diary, field notes and journaling

Useful precedents exist in literary authorship for the keeping of a diary or journal as a resource for ideas and later development of both style and of questions to be answered later. Examples of this journaling are available to writers today from certain authors who are studied for their diary-keeping practices, for example, Franz Kafka and André Gide, to mention only two. In Kafka's diary from 23 July 1914, he details a scene from a space of leisure in Berlin. Kafka focusses on a specific character working in the street and enters into a state of enquiry that simultaneously populates the everyday space and begins his interrogation of the urban space. Consider, too, that Kafka was journaling for himself as a literary writer attempting to develop his methods and was not keeping these notes to be read in this state by his audience:

> Evening alone on a bench on Unter den Linden. Stomach-ache. Sad looking ticket-seller. Stood in front of people, shuffled the tickets in his hands, and you could only get rid of him by buying one. Did his job properly in spite of all his apparent clumsiness – on a full-time job of this kind you can't keep jumping around; he must also try to remember people's faces. When I see people of this kind I always think: How did he get into this job, how much does he make, where will he be tomorrow, what awaits him in his old age, where does he live, in what corner does he stretch out his arms before going to sleep, could I do his job, how should I feel about it?
>
> *(Kafka, 1988, 293)*

Looking a little deeper at the tenses of the verbs, they shift from non-existent in the first three, note-like establishing shots, next, into past tenses as he describes the ticket seller's practices and then moves to a present tense and poses questions to test his own values and emotion on this way of living. This extract from Kafka's journaling in Berlin provides a useful lesson on the axiology of everyday practices and how they can be recorded as field notes for later use. Since Kafka is sitting securely on a bench, a stable, flat surface, he might be at work writing the field notes as the scene unfolds before him. It might be a few moments after, though, because he has had sufficient time to form an axiological analysis of the ticket seller's job and his performance or skill at practising it. Kafka has noted that the clumsiness is not impeding the seller's performance. This is a very minute synthesis of knowledge to share with readers. It fits the category of a small story (Patron, 2020), since it leaves a very slight trace of value. Arguably, if Kafka had not captured and then made this synthesis close in time and space to the witnessed scene, he would probably have lost it. It is a lesson in capturing these moments, along with any synthesis developed and tested on the spot to bring back from the field when travel writing. Kafka's diary entry captured here on 23 July 1914 leaves him a couple of personal knowledge management challenges: When will he answer those questions he poses himself? And, how will he use the small story in something he will publish for readers of his literary writing? It seems a memorable scene when isolated and examined here, but Kafka's diary moves to a meal with E. in the Restaurant Belvedere on the Strahlau Bridge, and just a little further on, to a fully-formed test piece of literary writing.

Four years earlier than this Berlin entry, back in 1910, a couple of weeks after Kafka's 27th birthday he makes a series of at least six starts at writing what seems like the reminiscences of an older man looking back on his life (Kafka, 1988, 14–21). Here, the journaling is not field notes, but repeated and modified tests of a piece of material about family relationships rather like sketch-booking in the visual arts. Since he had no word processor, he leaves every iteration of the design for later readers, even though this was unintended on his part. Each paragraph starts with almost the same reflective opening sentence, 'When I think about it, I must say that my education has done me great harm in some respects' (Kafka, 1988, 15). The second attempt uses that same sentence, then the third shifts slightly to 'Often I think it over and then I always have to say that my education has done me great harm in some ways' (Kafka, 1988, 15). These attempts grow longer as he writes them out, and soon his family begin to appear from his memory. It is towards them that he directs the reproach for this education, so that at one stage of the test pieces he lists the people he blames as if he sees them standing together in an old family photograph:

> Among them are my parents, several relatives, several teachers, a certain particular cook, several girls at dancing school, several visitors to our house in earlier times, several writers, a swimming teacher, a ticket-seller, a school inspector, then [...]
>
> *(Kafka, 1988, 15)*

Out of this writing practice, two ideas emerge. First, the care of subtly shifting self-identity that the literary writer has at their disposal, that will help them design and choose the persona they need to communicate a particular place. And, secondly, the conceit of reflection which does not use the more obvious mirror, but instead the idea of an old photograph to help create a background of memories that readers will understand because it is in their knowledge, too. In place writing, this same effect can be reached with the contemplation of an old holiday postcard of one of the shops or monuments in the town.

Finally, Kafka's notes can be seen being used as an inspiring catalyst for future travel writing, because of his canonical status as an author with translations available in other languages than the one he chose for his career. In a diary entry for 6th April 1917, Kafka recounts a dialogue he has with a workman at a tiny harbour. Kafka is curious about a 'clumsy old craft [...] masts disproportionately tall [... with] yellowish-brown sails' (Kafka, 1988, 373) and is told that it only puts into the harbour every two or three years and is the ship of the Hunter Gracchus. W G Sebald takes up this small story and weaves it into his travel book, *Vertigo*, in the chapter on Dr K at Riva (Sebald, 2002, 163). Thus, for present-day tourists staying on Lake Garda, they can take along with them both Kafka and Sebald as they explore the harbourside. These points from Kafka's diary demonstrate how the experimentation with identity, memory and narrative can be stored to be developed later during the write-up of a final literary text. Storing, indexing and retrieving such complex narrative knowledge presents a demanding task for the travel writer; in Chapter 5, the use of easily available technologies and conceptual systems, for example, the zettelkasten, will be explained.

## Theories of memory for writing enquiries

Bartlett's experiments with memory from the 1920s, as explained in Wagoner, Brescó & Awad (2019, 16–18), furnish a ready-made method for re-telling experiences from memory by case-participants or by the travel writer-researchers themselves. Wagoner et al. (2019) show how the re-teller makes qualitative changes as the experience is re-told. It suggests that memory is not a storage archive, like a computer, but is transformative, in that it integrates the re-teller's own culture as it is mediated appropriately for the next group of listeners or readers. The re-teller is thus an active agent, which, for the work of the heritage interpreter, can be considered an enrichment of the recalled experience. The term for this type of remembering during re-telling is called Bartlett's ghost memory, since the re-teller's own culture is overlaid on the original story. If some component of the original culture is not understood, then the re-teller recounts a ghost of that component, modified to be comprehensible in their own culture.

Keith Ansell-Pearson (2010, 161) proposes a reading of Deleuze where the creative process of art-making, and that includes literary texts, requires the overcoming of memory; 'whenever we think we are producing memories, we are, in fact, engaged in 'becomings''. Ansell-Pearson uses this position to shed more light on Marcel Proust's involuntary memory, and Proust's descriptions of the resort of Balbec, based on the seaside town of Cabourg on the Normandy coast. Proust's madeleine cake moment is well-known; two other sensuous experiences are also recorded by Proust as catalysts for involuntary memories, the stiffness of a napkin, and uneven cobblestones.

Each fragment of sensual memory, when it returns 'provides the narrator with the only genuine pleasures he has known, which are deemed by him to be far superior to social pleasures' (Ansell-Pearson, 2010, 163). It is these recollections that the narrator in Proust deems to be the truth of the past, and these are the experiences that make him want to become a writer to recount these. The recollections though do not describe the places, Balbec, for example, but rather tell of the essence of life's experiences there.

The third theory of remembering places is Sartre's development of the concept of the imaginary (Sartre, 2010). It is a noun from the French word, *l'imaginaire*, so it creates a moment's hesitation for English-speakers because in English, it is most often used as an adjective. Sartre explains the imaginary by asking readers, who have been to the Panthéon, to conjure up an image of this landmark building in Paris, the columns, the triangular pediment and the raised dome. The image is clear. Now, the challenge comes, Sartre asks, 'How many columns does the Panthéon have?' Try as they might, they cannot move around in their minds to interrogate the monument and count the columns; Sartre analyses this impossibility and draws the following conclusion:

> Thus we are led by this analysis to recognize that the space as imaged has a much more qualitative character than does the area of perception: every spatial determination of an object as imaged is presented as an absolute property.
>
> *(Sartre, 2010, 128)*

The researcher's imaginary holds complete but qualitative images of buildings and spaces that they know well, for example, the façade of their school building, or famous landmark buildings in the tourism spaces of cities they have often visited, and these qualities can be recalled and re-told in their travel stories. However, if a quantitative description is needed in the recounting, then the number of pillars or steps or windows is impossible to remember.

Can these three theories of memory help to enrich the work of the literary travel writer in the field? All three contribute to the changing identity or the becoming of the writer. Being aware of Bartlett's ghost memory concept allows the writer to gather stories in the field but with awareness that they are enveloping these in their own or their readers' culture. Proust's madeleine moments remind the writer to engage more sensually with the tourism spaces they visit. The writer in the field can stop and taste local drinks and food, sense the gravity of climbing steps or uneven paving, catch the aroma of local flowering plants and feel the changing air temperatures as evening descends, and document these in field notes to link them to the emotions in their writing. Finally, Sartre's notion of the imaginary, encourages the travel writer to count those architectural features that readers will find of value in the later story. For example, how many pillars support the Chehel Sotoun pavilion in The Persian Garden in Isfahan?

## Memory and journaling

Building a strong journaling process will support and later activate the writer's memory. Initially, in the experimental work for this book, it was found that the discipline of journaling was very demanding since it took the working researchers into new, untried practices of note-keeping and note-taking from other texts. If a researcher has been working for several years before trying to build a new journaling practice, then it is worth looking at how the researcher keeps and finds older material that they have previously published. Academic researchers might store their published papers on a Web 2.0 social platform, for example, ResearchGate, or as part of their own institution's repository on a PEARL system. First, then, it is worth analysing if they return to those sources as they begin to write a new piece. Indeed, it is only the need to write a new piece that stimulates the note-making whereas this research is proposing that journaling takes place practically all the time.

The primary design consideration then from this, for a new journaling practice and its support system, is to make notebooks available in as many places as possible for the researcher-writer. Having one single, precious notebook may mean that the researcher does not want to spoil it with odd ideas and that often it is not nearby, but in the office or downstairs. So, a stock of half a dozen similar notebooks that will lie flat, provide hardback support for writing while standing and have thick enough paper to resist bleed-through is a good starting point. Then, distributing these around all the places that the researcher is likely to be, along with pens, will enable new ideas to be captured, field notes to be made and more extended, what this research calls, affirmation journaling to take place. Thanks to lens software apps, these handwritten notes can be scanned easily to convert them to digital text for word-processing.

The notebooks themselves will not be marked out with any pre-printed template but during the work for this book, two rows of a table were designed in which to store these initial note-taking ideas and moments just a short time after writing and scanning them. The design in a word processor table is this, Table 4.1, although the table can be created in Microsoft One-Note as a template for rapid re-use.

The top row, labelled with the open inverted commas icon is to help capture the catalyst that made the idea seem valuable enough to write down and keep. In practice, it was found that if ideas are left even a few hours, then their value seems to wane or they become disconnected with the instigating idea. In the long term, if the prompts or catalysts are kept, then the researcher can return to the most stimulating of them on new projects when new ideas are not forthcoming. A catalyst can seem to come out of the blue, but if the source can be held onto, then here is a table cell in which to keep them. Research questions do, of course, prompt responses and ideas and so these are represented by the word 'Enquiry' in the label cell. To be multipurpose, though, this is also the place where the researcher will paste-in

**TABLE 4.1** Zettelkasten journal template design, detail of first two cells

| | |
|---|---|
| <br>Enquiry, quote or theme | *Catalyst for this idea:*<br>Letter, Mina Harker to Lucy Westenra.<br>'17 September.<br>My dearest Lucy,<br>So here we are, installed in this beautiful old house, and from both my bedroom and the drawing-room I can see the great elms of the cathedral close, with their great black stems standing out against the old yellow stone of the cathedral and I can hear the rooks overhead cawing and cawing and chattering and gossiping all day, after the manner of rooks—and humans [...]' (Stoker, 1897, eBook not numbered) |
| <br>Affirmation journaling | These views that she describes from the two different rooms might help locate the house from where Mina is writing, and the orientation of the house. Are those old elm trees still in Exeter in 2022, now older by another 130 years? Even if not, their mention in the novel makes them worth searching for during the fieldwork |

Designed during the research for this book (Mansfield & Potočnik Topler, 2022).

quotations from texts they are currently reading. Quotations or the paraphrase of another author's argument often sets research in motion; it is a miniature literature review. Finally, the label 'Theme' helps researchers transcribe and store structures of complexity from their mind-map or brainstorming sessions. One example in the work for this was the inclusion of a botany element and how plants are encountered in the urban space.

The notion of affirmation in journaling is drawn from posthumanism, which seeks to find development and positive growth from the ideas. It sits alongside criticality, which is traditionally used in reviewing academic texts, but creates openings for writing-out ideas to find positive ways forward. For further reading on this, Rosi Braidotti devotes a chapter to affirmative ethics in Braidotti (2019, 151–173). The scanned handwriting from the researcher's hardback notebooks can be converted into digital text and then pasted directly into this lower of the two table cells. Sometimes, this might be the first cell to be used, and the catalyst cell above completed later. In the research, the photographed scan of the original handwriting is also stored, associated with the digital forms. This means that the digital text can be edited without the original being lost, thus its genetics are archived.

Editing and re-working the digital text in the affirmation journaling cell is an important part of the processual practice put forward here. This text should be brought up to public-facing quality. First, because it creates a resource for the researcher that can be used almost immediately as a pasteable paragraph for any final, published work, and this hands back value to the digital version of the researcher's journal. If it holds usable texts, then it is worth keeping and going back to. Second, which is a hard discipline to learn in the early stages of journaling, is that it is a sandpit or workshop to which the researcher-writer will return to re-work and develop ideas to become more substantial and substantiated. When experienced publishing academics have to make this change to work inside the online journaling environment, rather than write their articles direct, this was the most difficult change to make.

To make this change more attractive, a further field or row was added to the online template, references. Here, the current books and journal articles can be converted to the preferred citation standard and then are always present when this journaling is re-visited or extracted for writing-up (Table 4.2).

Additionally, up to five other fields are added to the digital version of the design for journaling, all of which provide clickable ways of cross-referencing this journaling page with other source material or leave keywords that can later be searched upon to re-discover this note. Finally, a space for dialogue keeps the note alive and open in the spirit of the anarchive (Table 4.3).

It is intended that dialogue can be opened using Web 2.0 to writers cooperating in the project, or the commissioning body, for example, the DMO, or to stakeholders with whom trust is being built, or a teacher in a learning situation. When the affirmation journaling note and the dialogue have matured, then the next step for this can be a blog post to a wider public. The digital process is looked at in more detail at the end of Chapter 5 in this book.

TABLE 4.2 Zettelkasten journal template design, detail of references cell

Stoker, B. (1897). *Dracula*. Edinburgh: Constable & Robinson. https://www.gutenberg.org/files/345/345-h/345-h.htm

References

Designed during the research for this book (Mansfield & Potočnik Topler 2022).

TABLE 4.3 Zettelkasten journal template design, detail of dialogue cell

Q: Do you know if and where elm trees still stand near the cathedral in Exeter?

Dialogue

Designed during the research for this book (Mansfield & Potočnik Topler, 2022).

# 5

# METHODOLOGIES AND PRACTICE

## Writing as a processual practice

This chapter places emphasis on writing as a process and as a processual practice, that is, as an activity with resulting output that can be changed at any time by the writer before publication. It is worth posing the question, 'is it necessary to go through all the processes discussed in this chapter when creating a piece of literary travel writing?'. This requires the following claims to justify the processes that will be covered. Treating a place-making or travel writing project as a process helps improve the repeatability of the work. This makes it more practical for the professional writer to provide quotes and delivery schedules for writing commissions.

1 A methodical process with comprehensible methods provides the writer-researcher with rigour of approach. In moving towards posthumanism from traditional ethnography, the community of peers who will read and use the final travel piece need reassurance that the production has gone through a rigorous process to ensure its credibility and usability by its audience. The travel story is a new piece of knowledge for those that share the writer's values.

2 One of the sub-groups of the audience will be the stakeholders of tourism in a destination. The process must be plausible as creator of value for the place and the people concerned with the value of this place. In this way, the process is both a method of inquiry to discover value in the place and a creative method to ensure that this is communicated sensitively and effectively to those interested. A documented process helps the researcher to see the steps taken and to note which are effective, both in the interrogation of space and in the output.

DOI: 10.4324/9781003178781-6

3 The developing identity of literary travel writer will be better understood if the steps in each project are made visible. In colloquial language, it can be seen what worked, what was successful in, say, communicating affect to the stakeholders. Through iteration, the skills of the travel writer can be improved and effectively professionalised since writers following a set of methodical steps can better predict and estimate project times, costs, requirements of access to people and places and potential impacts of their work from maintaining historical records. This also includes emotion and affect, as an essential part of literary communication.

In literary writing, design points are often applied by writers to develop their own range of skills and to experiment with the writing to test design methods. For example, a writer might decide to focus on the plants and botany of a destination and even plot a route around the town that follows the outline of a tree or a leaf to communicate the idea of the plant to readers. Or, the focus might be on a moment in history when a major change took place, for example, when in 1680 a Custom House was built on Exeter Quay to tax woollen cloth exports. A list of these design points can be made as part of the planning process but this does not mean that they must all be included. The length of the final travel story will in part determine how many design points to include. Traditional guide books can provide building by building descriptions of architectural features of every part of a town's built heritage but the literary travel text might only consider the structure of one façade and make emotional connections with the geology or with the economic changes taking place at that moment in history. A longer list of design methods, the D-methods, are given in Mansfield (Mansfield, 2020); it must be stressed again, though, that not all these D-methods need to be included in every piece of literary travel writing. They should be taken as catalysts for experimentation and then a decision taken whether to include the passage of writing emanating from that experiment. Thus, they are part of the processual nature of the writing inquiry.

### Introduction to practice as research

As postgraduate researchers begin their travel writing projects at Masters, and more particularly at doctoral level, they try to apply the methodology of the reflective practitioner to progress their new writing. This often stalls the beginning of new work. Three key reasons create this block. The reflective practice approach requires an existing productive practice on which to reflect, so that for example, the experienced historical writer, Hilary Mantel, advanced in her published career, can productively reflect on a day's writing. In podcasts and radio interviews, Mantel talks about how she aims to use real quotations of language from historical sources to give voices to her characters. Mantel has a practice that she has identified and shows her reflection on this in public interviews, whereas the researcher embarking on a new practice is by the nature of researching, still in no position to reflect on an established productive way of working.

The other two reasons why reflective practice fails to launch a new literary travel writer is that the reflective approach comes from school teaching. At the end of the teaching day, the primary school teacher, who has dealt with the needs of 30 pupils and several colleagues and parents, has a wealth of social interactions upon which to reflect. Further, school education has a framework set by government and a method of assessing the effectiveness of the demands of government in the form of national testing. Hence, the reflective practitioner in education has a set of metrics against which to reflect. The new writer does not have these surrounding requirements, nor a day's productive practice behind them.

If the travel writer turns to frameworks and benchmarks for teaching creative writing, the usual approach found in degree programmes for this is workshopping, where the day's writing is shared with a peer-group for a group critique (QAA, 2019, 10 *et passim*). However, literary travel writers, while benefiting from comments on the effects of their writing, still need a plan for developing their practice and metrics they can apply to assess progress towards some goal. The model might best be conceived of (i) plan from aims, (ii) practise the writing and (iii) evaluate against the plan. This is a very simple process taken from software design, but it does move towards practice-led research, particularly if the planning is considered as part of the practice rather than the writing alone being the practice.

Because travel writing emerges from tourism management teaching within business schools or human geography departments, it is informed by (a) social science methods and (b) the design process in writing but posthumanism seems to open up more productive ways of doing and making in travel writing. The Creative Writing benchmark statement, the document the QAA provides for every subject taught at university, introduces different ways of conceiving knowledge from the business and management benchmarks. In particular, for this chapter on research methodologies, the following six numbered points from the list of 17 subject knowledge points expected of a creative writing graduate help form a plan both for the literature review but more importantly here, for establishing a relationship to the discursive practices surrounding the destination under study:

> Creative writing graduates have subject knowledge in:
> **v** the creative process and the body of written works that surrounds it
> **ix** the relationships and interactions between different genres and media
> **xiii** theoretical, generic and practice-based concepts and terminology
> **xiv** critical awareness of the context in which writing is produced and how individual practice relates to that of predecessors and contemporaries, peers and established practitioners

xv the nuanced critical awareness gained from contextualising their own writing within a given framework, historical, cultural or generic

xvi the development of new writing strategies drawn from critical reflection upon their writing practice.

*(QAA, 2019, 6)*

The literary travel writer will seek out and use methodical processes for dealing with technical, trade and scientific texts and spoken or mediatised discourses that will build a framework around the destination of study. This framework building responds to the six points above, most specifically to v, xiv and xv. From understanding shellfish habitat, geomorphology of seaside harbours and public tendering rules right through to calculating power consumption in kilowatt hours, each new urban environment under study will present a new set of transdisciplinary knowledges for the travel writer, as this chapter shows below.

### Writing in the public sphere as research

Social media, especially online web-logging software, including WordPress and Google Blogger, provide a more immediate dissemination of research ideas and experiments than traditional routes of publishing. Where travel writers are also researchers the writing can also be deployed as an instrument of inquiry. At least two processes are taking place when publishing work-in-progress to a public, social space. The first is the effect felt by the writer, using the concept of face-to-face rapport from Lévinas (Lévinas, 1985), and further extended by Bakhtin's idea of answerability (Nielsen, 2012), the writer is in an ethically aware state that others might be reading this output. The second process is the impact on the readers and the various levels of engaged feedback they might provide to the researcher.

The primary, very direct feedback mechanism is to set the blog software to accept readers' comments or to align the blog with a Twitter feed so that readers do have a method of answering blog posts. This feedback, along with focus groups or email questions, can provide the researcher with open direct readings of their work. Referring a respondent to a blog link, the url or web address and asking direct questions in an email, though, allows a more elaborated level of research inquiry using the writing. This level can be seen as three distinct areas: forming impact bonds through networking, building gatekeeper trust for more detailed later inquiry, and citizen science, where a very large audience may hold one or two highly specialist respondents. All of these groups develop the writer's knowing through narrative accumulation.

A valuable preparatory step before travelling to a new destination for fieldwork is to conduct a literary geography. This is to uncover and find novels and other literary and emotive texts written in or set in discernible locations within or en route to the destination city. In December 2020, the project 'Be Wallonia' used this preparatory method of uncovering the literary geography through the author's own blog, and very shortly after the start of the blogging via the blog of the destination management office, the DMO of the Wallonia Foreign Trade & Investment Agency in Brussels (Mansfield, 2020). Building trust with the DMO resulted in wider networking possibilities, opening the gate to other researchers, and to other organisations who would contribute to the knowing of the researcher. In citizen science terms, the blogging coupled with a Twitter feed led to the discovery of novels set in the towns of the researched region, for example, Georges Simenon's (1931) detective novel, *Le Pendu de Saint-Pholien*. The interaction established with an engaged and highly knowledgeable respondent audience who were already in the field gave rich data on the location of a building in the area of the town of Liège, alluded to in the book's title. Reading the novels, although enjoyable, does lengthen the time of the fieldwork preparation; this use of blogging as a tool of inquiry helps pinpoint texts that unfold in knowable streets and places in the cities to be visited.

### Designing the blog to maintain momentum

Using a blog as a research store for a specific and constrained period of time within a travel project or place inquiry is a useful introduction to the discipline of regular writing production of public-facing text. Deciding upon and designing the frequency of posts offers writers their first exercise in structure. It is also the first trial of how much and how often they can write. For example, a week-long journey can be structured by days of the week, as in Rötig's (2018) *Cargo*. The posts could then be made every evening to recount the day's findings. A more theoretical starting point is taken by Antonio Nobile is his web log called The Unknown Town. Concerned with achieving a complete piece of work, Nobile uses a finite list of elements from a typology of travel texts (Table 5.1).

The anatomy provides the student writer with the discipline to find material for, and then write on each theme at certain times during the planning and the fieldwork. This range of themes supports Nobile's quest of finding the unknown town since it explores the town using approaches revealed in other travel texts. More targeted typologies of themes can be brought into use as the calendar of posts if the quest is clear or taken from a brief. One example is the use of a list of themes for exploring ethnobotany. It must be remembered, though, that not every element of the typology or table must be found; it is part of the design and the aims for that particular text that will help determine the amount of points that are explored (Table 5.2).

In a series of experiments with the Travel Writers Online blog in 2020, the length of a post was tested. The results found that handwritten A4 pages held 350 words; of course, this will depend on each writer's style. These 350 word pages were taken as a baseline of sustained writing from one sitting of between 30 minutes and 45 minutes. Three pages, that is, 1,050 words,

**TABLE 5.1** Anatomy of travel texts

|   | *Theme* | *Analysis* |
|---|---------|-----------|
| A | Anticipation | Mounting excitement at the prospect of the journey, but also see xéniteia below |
| B | Books | Preparatory reading |
| C | Clothing | Identity shifts possible in new clothing at the destination |
| D | Displacement | Displacement and time are components of travel movement so verb tenses will provide inroad to textual practice |
| E | Episteme | The travel text will add to the stock of knowledge |
| F | Food and drink, meal-taking | Strange new foods. Meals prepared by someone else. A pause in the journey is invested with more value if eating or drinking |
| I | Images, sights | The travel writer will see new and beautiful things, for example, views of Paris as a picture postcard |
| L | Language | The strange language may not appear connected to the travel writer's own world. Writer may choose to incorporate found texts, spoken or written |
| M | Map | Printed page will use white space as part of structure of travel text, reminiscent of the map |
| R | Responsibility shift | Traveller is at ease, responsibility seems removed allowing traveller to behave outside home conventions |
| S | Self | Self-identity inscribed in the text as exote but entropy may be at work |
| T | Transport | Mode of transport contributes to literariness of text |
| U | Uncanny | Sights and new people will recall previous literary or artistic readings |
| V | Veracity | The travel writer will report the 'truth' as they see it at the time through their culture |
| W | Weather | Used to render simultaneously truthfulness and literariness |
| X | Xéniteia | Deciding what to take with you on the voyage and what to leave behind. Putting affairs in order to live an organised life |

Developed after Mansfield (2012, 75).

was sufficient to form a three-page feuilleton. This three-page structure gives writers two moments in the whole text where they or their readers can pause, between p. 1 and p. 2 and between p. 2 and the final page. This creates a challenge for writers to provide an encouragement for their readers to continue onto the next page. Variously, this can be called a cliffhanger, a teaser, a trailer, a question or a page-turner. From this, the blog posts can be published in a way that will encourage the readership to follow the calendar

**TABLE 5.2** The ethnobotany checklist for literary travel writers

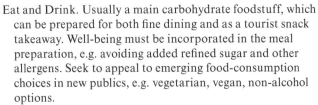

*The ethnobotany checklist*

 Eat and Drink. Usually a main carbohydrate foodstuff, which can be prepared for both fine dining and as a tourist snack takeaway. Well-being must be incorporated in the meal preparation, e.g. avoiding added refined sugar and other allergens. Seek to appeal to emerging food-consumption choices in new publics, e.g. vegetarian, vegan, non-alcohol options.

 Food for pollinators to maintain the plant and botanic life forms. These plants will require education panels and links to local symbolic cultural capital.

 Recyclable plant products. Raised social awareness of whose labour and knowledge is used and is valued in the collection, processing and re-use of the plant by-products.

 Food taste or scent – a local herb, dating from ancient times, perhaps held in a local place-name or the name of a river. Symbolic – positive with local connection to culture and literature.

 Apparel. Wearable botanics, created and made with local craft skills. Institutions for this knowledge to be preserved, enhanced and transmitted as part of local culture in colleges, workshops and through courses offered by regional universities.

 Shelter – building with botanics and using local production knowledge and labour. Visitor attraction, equipment or building made of wood. E.g. old wine press, bench, a tree park or arboretum. Wooden buildings using local timber rather than imported timber. Sponsorship and promotion of local timber and forestry businesses brought into the tourism space.

Designed during the research for this book (Mansfield & Potočnik Topler, 2022).

set by the blogger. This is important for time-sensitive work where writers are hoping for near-live dialogue to enrich what they know of the place and the route. In a supported learning situation, e.g. university, the teacher or PhD supervisor has the opportunity to post live feedback to the learner during the blogging process. This focus on timing the production of texts with known word counts is valuable to professional writers since it provides them with real baseline data for planning their own time in the field. For commissioned work, it will assist in providing time and cost estimates for clients and printers. For those using this book to support their postgraduate research at Masters or doctoral level, their having a realistic figure for writing productivity will help address project schedules. Typically, the long dissertation module at Masters in the United Kingdom is now a 60-credit module, that is, 30 ECTS, with a written length of 15,000 words. A PhD thesis in the United Kingdom must not be longer than 80,000 words, and this includes any appendices. Any longer than this requires agreement from the external examiner, and a further fee payment. In practice, doctoral candidates aim for 75,000 words; this count does not include the list of references, which should be around 300 books and journal articles as a guide.

### Thesis chapters 75,000 words

The chapters of a practice-led literary travel writing PhD could have this pattern of word counts as an initial framework to help the doctoral researcher produce a project plan for their research proposal (Table 5.3).

Own Practice – Chapter 7 in the thesis is an opportunity to present the researcher's own travel writing and place-making texts. An example of the types of writing that can be presented to the examiners is given below, labelled A to G:

A  A literary geography of a place, e.g. a town nearby or in the region or en route.
B  A hexis plan with an inquiry question – see Cherbourg Chapter 6 in *Cities*.
C  Field notes for six blog posts, taken live in the field, aim for one-side of A4 lined, circa 350 words per post. Use this to learn how many words you write per A4 page. How many words you write in a half-hour. How easy and quick it is to write. Do you repeat words?

**TABLE 5.3** Chapter word-counts for a travel writing PhD project

| Chapter | Word count |
|---|---|
| 1  Introduction & Aims<br>Literary is defined, research questions set as aims | 4,500 |
| 2.1  Literature review of literary travel writers and their practices | 8,000 |
| 2.2  Lit. Review of Writing as Place Inquiry and Practice-led research with writing | 8,000 |
| 3  Conceptual Framework along with more on the axiology, stakeholders and audience of this writing | 7,000 |
| 4  Research Methodology and Research Instruments. Data collection and fieldwork documented or proposed. | 9,500 |
| 5  Data and findings. | 8,000 |
| 6  Analysis and synthesis | 8,000 |
| 7  Own practice | 15,000 |
| 8  Conclusion & implications | 7,000 |
| Total | 75,000 |

Designed during the research for this book, (Mansfield & Potočnik Topler, 2022).

D The six posts blogged as two three-part feuilletons, and Comments elicited from readers and followers.

E A feature-length article written and pitched for a print travel magazine or travel section, with 1,500 words.

F The major literary travel writing on the fieldwork. Consider as a starting point, 55,000 words modelled initially on Sven Lindqvist's 150 entries of 350–370 words in his travel book, *Exterminate all the Brutes* (1996 EN, 1992 Original Swedish).

G Personal Journaling as data, may be handwritten and only scan or photograph small sections when you want to paste them in for supervisors or into thesis.

Impact Indicators – these demonstrate how the completed texts have reached an audience and the affect these have had.

i Demonstrate engagement through, for example, sponsorship, through blogging, e.g. Notebooks, Apps used, apparel, stakeholders.

ii DMO involvement, which might be a commission or recorded meetings.

iii Magazine editor or travel pages editor.

iv Correspondence with book publisher, or literary agent for full-length literary travel book.

It is helpful to have in mind the QAA's descriptor for a higher education qualification at level 8, that is a doctoral level degree, when planning thesis chapters and their contents since the submitted thesis must demonstrate to external examiners these points:

Doctoral degrees are awarded to students who have demonstrated:
- the creation and interpretation of new knowledge, through original research or other advanced scholarship, of a quality to satisfy peer review, extend the forefront of the discipline and merit publication

- a systematic acquisition and understanding of a substantial body of knowledge which is at the forefront of an academic discipline or area of professional practice

- the general ability to conceptualise, design and implement a project for the generation of new knowledge, applications or understanding at the forefront of the discipline, and to adjust the project design in the light of unforeseen problems

- a detailed understanding of applicable techniques for research and advanced academic enquiry.

Typically, holders of the qualification will be able to:
- make informed judgements on complex issues in specialist fields, often in the absence of complete data, and be able to communicate their ideas and conclusions clearly and effectively to specialist and non-specialist audiences

- continue to undertake pure or applied research and development at an advanced level, contributing substantially to the development of new techniques, ideas or approaches.

And holders will have:

- the qualities and transferable skills necessary for employment requiring the exercise of personal responsibility and largely autonomous initiative in complex and unpredictable situations, in professional or equivalent environments.

*QAA (2014, 30)*

### Preparing for the field, themes and axiology – value, and value to whom?

The transdisciplinary field of inquiry when travel writing is the research instrument usually leads to rejection by established academic journals because the research work is outside the scope of traditional discipline areas. For example, in heritage studies, the other researchers in the field are looking for a contribution to understanding a heritage artefact or one of the key issues in their discipline; travel writing needs to approach that subject then rather than reflect upon itself. Which leads to the question, what can travel writing examine? Three key themes emerge initially: (i) the writer's own identity, (ii) place inquiry including the heritage artefacts mentioned earlier and (iii) more traditionally, the practices of the peoples and culture encountered by the researcher in the field.

How are any of these three themes of value to the researcher, to the discourse practice peer-group, e.g. journal editors, or to the public, either a general readership via commercial publishers or as impact on public policy? Considering the value of the research and practice is part of the axiology of the project. If it is the writer's own identity that the researcher is valuing, then the choice of research will be an investigation that interests the researcher. Ethically, the work could be asking, what can I do? What am I capable of doing? However, this personal drive can be harnessed to a quest that will take the researcher out on a journey which gives the investigation value to a public readership. An example of this is the travel book, *Cargo* (2018) in which Rötig connects her own fascination with container ships to an exploratory journey starting with train travel to Le Havre.

This personal interest, with a journey added, could be further enriched from an axiological point of view by exploring social and economic changes. For example, in early 2022, public consultation took place in port towns across Normandy, in preparation for the construction of a wind farm in the Bay of the Seine. With 150 companies and the French government involved, the aim is to reach a generating capacity of 1.5 gigawatts of electricity within the 2020s. What can a literary travel writer do with this change in the way electricity is generated? One fishing village of 559 inhabitants, Barfleur, is set to be the closest harbour to the construction site. The population has been falling since the 1830s and the railway station closed in 1950. Already, it is apparent from this brief overview that a transdisciplinary approach is needed, suiting the research perfectly to the informed processes of a writer of narrative non-fiction. The writer will need to understand and find a way to explain what 1.5 GW means. Do visitors to Barfleur experience electricity? What is the current tourism and hospitality experience in the village of Barfleur? The public consultation process produced commissioned reports on impacts on wildlife and environment; these need skilful interpretation. And the emotions of particular villagers, who might be representative of a larger demographic, would need eliciting and recounting sensitively. The collection of all these documents is closely aligned to sketchbooking and in this way becomes a processual research practice. For the writer, this collection must be accessible and act as a catalyst for new approaches whilst for the marker, assessing the work in a university setting, the collection must also demonstrate the steps of synthesis that the writer took in their process and decision-making. For example, how did the writer decide on Barfleur? How has the writer worked with the French term for a large-scale wind farm, Aeolian Parc, to uncover any literary affect?

Alongside desk-based reading of the documents, and once the initial journaling has made a particular destination have value across the personal and environmental realm, then an engagement in dialogue with a range of others will enable the writer to maintain an observatory on the axiology of the project. For instance, dialogue with a cooperating writer or simply a friend or colleague who has no stake in the work can reveal where, in the collected notes, any public interest lies. This process is a narrative dialogic process. The writer endeavours to articulate the work in progress, and, if colleagues show interest, then they will ask why an element has been included, thus calling upon the writer to explain more. This process is narrated in the short novel, or sotie, by André Gide (1895) *Paludes*. Social media groups of specialists, e.g. ResearchGate, are a good source for finding very particular researchers who are currently active close to the studied area. An example of this in practice is on the Be Wallonia blog, mentioned above, which engages with the research of Julie Masset and Alain Decrop, who asked tourists where they placed their souvenirs when they return home (Masset & Decrop, 2020) and then reported their findings to show that many items brought back from holiday found their way upstairs to bedrooms, to wardrobes and even to the attic. This journey through vertical space of the objects into the intimate recesses of holidaymakers' homes provided an affective addition to the literary blog posts on topology and desire paths in the Belgian DMO's place branding.

If stakeholders can be found for this dialogue, then trust-building is initiated at this stage in the processual approach. These stakeholders, and it may be easiest at first, to contact press offices of the public bodies involved, will answer technical inquiries more readily at a later stage. The stakeholders might also become those affected by the final literary travel writing, and so be the site of impact of this research. In trying to make a living from the writing, they might also become commissioners or sponsoring funders for the continued research.

### The writing equipment, materials and documents

The diary format has a long pedigree as a literary form. Sartre's (1938) *La Nausée* [*Nausea*, 1963 EN] uses days, dates and times as side-headings throughout the novel, and references are made to the construction site of the new railway station in Le Havre and the old station to provide readers with a historical point, and elapsed time for the narrator. The new station was completed in 1932. At 10:30, for example, the narrator of the diary looks down from his room, which faces northeast onto the construction site of the new station and can see the Railway-workers' Bar:

> 10.30 [...] From my window I can see the red and white flame of the Rendez-vous des Cheminots at the corner of the boulevard Victor-Noir. The Paris train has just come in. People are coming out of the old station and dispersing in the streets.
>
> *(Sartre, 1963, 7)*

The narrator gives the readers just enough information to let them know that the old station is still in use for the arrivals from Paris. Later, he maintains a time frame for his walk from the bar:

> It's half past seven. I'm not hungry [...] An icy wind is blowing [...] on the right-hand pavement, a gaseous mass, grey with streaks of fire, is making a noise like rattling shells: this is the old station. Its presence has fertilized the first hundred yards of the boulevard Noir – from the boulevard de la Redoute to the rue Paradis – has spawned a dozen street lamps there and, side by side, four cafés, the Rendez-vous des Cheminots and three others.
>
> *(Sartre, 1963, 35)*

In fieldwork for travel writing, Sartre's format offers a good template to record dates, times, weather, street-names, and the personal feelings of the diarist. It is no coincidence that this is an existential text and it is the existence of the travel writer in specific locations that field notes aim to record. The writer-researcher aims to leave clues to themselves which may not have any apparent use at the time of handwriting the testimony of being there but may prove useful later during the creation of the artefact of the literary travel writing. Indeed, in Sebald's *Campo Santo* (2006, 44), the literary travel writer is back in his hotel room in Piana, Corsica. He tells his readers that he finds in the drawer of the bedside table a copy of Flaubert's story of St Julian, giving it even more solidity by mentioning that it is in the expensively bound Pléiade edition. This activity activates the place while Sebald is in the field. In fieldwork, the literary travel writer does activities to open and unlock the archive of the place. It is a practice drawn from the posthumanist process, anarchiving, which creates new texts that might be useful later. Besides unlocking, this anarchiving leaves recorded diffractions for readers to discover and unlock in turn during their reading of any subsequent write-up.

Here is an example of that from Jasna's field notes, which were handwritten in a beach bar. Jasna's use of the present tense for most of the written notes places her there but provides a time clue to when she was writing, since the order for the soft drink is written in a past tense. Along with the dateline and smaller time clues, which will be very valuable during any subsequent write-up of a travel narrative, a place clue also exists. First, though, here are the field notes transcribed:

> 18th July 2021, Tivat, Montenegro [...] It is 12:55 and I sit on the beach near the Tivat airport. It is very windy and nobody is in the water – we are all sitting in the Orange Beach Bar – around 10 guests – I ordered a Cockta a drink similar to Coca Cola but – so it is believed – more healthy and invented in Slovenia in 1953.
>
> *(Jasna, 2021, Montenegro Field Notes)*

The place clue comes from the re-use of the present tense of, to sit. 'I sit on the beach' but also 'we are [...] sitting in the [...] Bar'. This suggests that the bar is laid out on the beach and is not a building a short distance away on paved ground.

## Grounded theory in writing as place inquiry

Synthesising new theories which are grounded in the data collected in the field has become a highly productive method in tourism studies, based on the new directions given to grounded theory by Kathy Charmaz (2014). Three steps characterise this method: elicit longer testimonies of data, analyse the stanzas of the testimony using gerund coding and form new hypotheses using memo writing.

Where are the data? For a travel writer when conducting place inquiry, a source of these testimonies can be the local citizens who are living through a major urban upheaval. The gerund coding of the transcribed interviews reveals the acts that the local respondents believe they are making during the period they are recounting. Additionally, other writers' blog posts can be treated as stanzas and analysed using the coding process. A two-column table is a practical method of breaking up the testimony into separate stanzas and adding the gerund coding alongside, like Table 5.4.

**TABLE 5.4** Two-column template for gerund coding for re-use in grounded theory research coding

| Gerund coding using actions words ending in –ing | Stanzas from the respondent's transcribed testimony |
| --- | --- |
| Discover*ing* self.<br><br>Document*ing* self.<br><br>Display*ing* own cultural capital | Friday 28 March 2014<br>'The world is a book and those who do not travel read only one page', this is probably my all-time favourite quote written by St Augustine. I want to travel. All my life I have wanted to travel. To discover new places, the remoter the better. It is only when travelling that I feel truly at peace with myself. Any journey is for me as much a journey of self-discovery as of discovery of an unknown environment. In just under two weeks, we leave for Concarneau. It is still too soon for me to have the butterflies in my stomach as one might have just before leaving on any trip, but excitement is mounting. This time, anticipation for travel is greater than usual. I would classify myself as a very spontaneous and last minute traveller. On the 3rd of January, I went to Paris, I booked the flight the day before!! This time planning and research have gone into organising the journey. I think this is the first time I've read a book set somewhere and then specifically planned to go and visit that place. |

Designed during the research for this book, (Mansfield & Potočnik Topler, 2022).

**TABLE 5.5** Memo-writing template for re-use in the post-coding phase of grounded theory projects

| Memo-writing – Creating the identity of the travel writer | Identity |
| --- | --- |
| Opening sentences are written in the first person establishing the cultural capital of the narrator. The notion that the narrator is prepared to display their own cultural capital, and make discoveries about the self is made clear early in the story. | |

Designed during the research for this book, (Mansfield & Potočnik Topler, 2022).

As ideas begin to crystallise from the coding, the researcher starts to open memo-writing boxes to capture and develop the ideas, and some layout design is added to the boxes as tables: a working title, a colour coding of a category word at a higher level to group the gerunds (Table 5.5).

Later, after having documented the idea in memo-writing format, the ideas are followed up from either the academic texts studied during the original literature review or, if not already discovered, then through additional academic reading. References and quotations are then added into the memo-writing box until sufficient evidence has been accumulated to write up the new synthesis of theory grounded in the interpreted data. An echo of this processual approach, which returns to an ongoing, developing text can be seen in the Zettelkasten journaling proposed in Chapter 4 of this book.

## Emergence and genealogy in change-of-land-use documents

In the preparatory desk-research phase before planning a visit to a destination, it is valuable to uncover any major changes that are taking place in land use. In the study of the city of Nantes, a major project of land use change took place at the start of the 2000s (Agueda, 2014), as the city DMO changed the use of a tobacco factory and other quayside warehouses into a tourism space. Changes of this scale in the built heritage leave significant traces in the lives of former employees in the town, and in other users of the erased social spaces or users of the newly created spaces. Change of land use is highly regulated in western Europe, and this process of planning consent generates a growing archive of documents. The researcher-writer can approach these documents using a methodological approach drawn from Foucault's idea of genealogy by Maria Tamboukou (Tamboukou, 1999).

The building of the Aeolian Parc, or wind farm, in the Bay of the Seine had already produced an archive of documents by the close of 2021, created initially by the French government for its tendering call, and later, by the organisations leading the public consultation, that took place in the affected towns of Normandy in early 2022. For example, this document was made readily and freely available to the public:

Haskoning DHV (2020). Update on the Current state of Knowledge on the Environmental Impacts of Offshore Wind Farms, Exeter: Haskoning DHV.

Along with the documents like the one above, a range of maps and photomontages were commissioned to show the impacts on the lives and the urban spaces bordering the marine bay of the Seine. Three potential themes for the travel writer emerged from this collection, by applying Tamboukou's genealogy methodology (Tamboukou, 1999). First, the report reviewed studies on the use made by edible crabs of the monopiles, where the new wind turbines are embedded in the sea floor; this is likely to increase the crab stocks for the fishing ports on the bay. Secondly, the planned building of an observatory for marine wind power was discussed; this will attract and encourage qualified specialists to live in the bay area, and, thirdly, the port most likely to be used as the maintenance point emerged. Judging from the cartography, this is very likely to be Barfleur, which may also be the site for the new observatory building. With these emerging themes, the writer-researcher could spend time investigating the town Barfleur, establishing respondents who might be affected by these three themes, and begin to collect interview data and spatial data in the town.

## Self-writing and processual identity

Discovering and applying design features in written texts within a practice-led methodology will provide a mechanistic, practical process for increasing the number of aspects a writer can explore. These will be examined later, but first, what changes occur in the writer's own identity as they practise? And, from an ethical point of view, what should they be doing for their own identity? Foucault (1983) looks at the changing function of self-writing and journaling between the classical Greek era and the first two centuries in the current western era. Foucault calls these journals, hypomnemata, and in the singular hypomnema. He makes it clear that for the Roman writers at this point, he chooses Plutarch (AD 46–AD 119) and Seneca the Younger (c. 4 BCE–AD 65), this journal-keeping was not a diary nor a forward planner, but rather had an ethical function. The notes and citations were for care of the self. Foucault explains (Foucault, 1983) that Seneca selects maxims to copy down that are useful to himself in his current circumstances. These quotations might even be from writers with whom Seneca normally disagrees, Seneca gives the example of Epicurus (341–270 BCE), because he claims some part of the quote for himself, which can be understood as building his changing identity for that day. Of course, the journal will become a mass of miscellaneous ideas always in process to keep time with the needs and preoccupations of the self on that day. However, both Seneca and Foucault propose that this deliberate heterogeneity does become unified in the two steps of personal writing in the hypomnema and in their re-reading. Yet, the self, the writer is not indoctrinated by their transcribed quotations, Seneca uses the metaphor of the bee gathering and digesting food, the ongoing results becoming the new body of the writer in 'tissue and in blood' (in vires et in sanguinem)' (Foucault, 1983, 420), after Seneca the Younger (c. 65), letter 84 to Lucilius, 'On Gathering Ideas'.

This historical examination of journaling is not to claim any authority from precedence but Foucault explains these historical practices of personal writing as part of his genealogy of the care of the self. Foucault detects a later subversion of the purposes of self-writing which leaves the uses espoused by Seneca in a dark fog, forgotten by subsequent keepers of notes. For contemporary readers, Seneca's practice is illuminating because of its freedom or openness to other authors, and through its processual nature, that is, a refusal to fix and fossilise the identity of the researcher-writer, ideas which are both found in posthumanist re-thinking in the twenty-first century. Rosi Braidotti is aware of the limitations of a researcher's current knowledge, but offers the posthuman term 'adequate understanding' (Braidotti, 2019, 131) which creates a power to act, to feel joy in this power and thus to gain a renewed desire to continue gathering ideas. For Braidotti, this joyfulness in searching for newer knowledge of complex assemblages is a component of the ethics of freedom after Spinoza.

Processual journaling, then, carefully places and even copies testimony and evidence into a heterogeneous body of work. The pieces of evidence are not a static archive but an opening up of the developmental steps, the choices, discoveries and decisions made during creation of the travel story that the researcher will go on to write from this journaling. Indeed, the writer might decide that the final travel text will leave some of the pieces of evidence unlinked as an invitation to readers to verify, geographically or historically, the account or to signpost an avenue not taken in this text but related to, and supporting, the main story. Walter Benjamin, in his advice to narrative writers, says the good storyteller resists too much explanation, leaving space for readers to engage in the story by providing their own thinking and reasoning (Benjamin, 2019).

The processual journal is proposed as a practice that goes further than the more traditional notion of a reflective journal. From a literary point of view, the metaphor of Narcissus gazing at his own reflection in the lake emerges when considering reflective practice. The researcher-writer will return to their early journaling as much as to their field notes when writing up the travel story. It is useful, therefore, to create page layouts that will allow space for returning comments or to map out a literary geography as a second step to planning the route of the field trip. For a literary geography, the researcher seeks out novels and diaries set in the destination under study. For example, Modiano's (1975) *Villa Triste* is set on the northern end of Lake Annecy. A sketched journaling page can explore whether it is worth designing a travel route to take in the journey around the lake to Veyrier-du-Lac, because a key moment in the novel's plot takes place at the Resserre restaurant in that smaller town. Using a sketch-booking approach to journaling, it is easier to see that a visit around the lake would be a journey of 7 kilometres and so would not be comfortable to add to an urban walking route. No restaurant of that name could be found, either. However, the idea of the three main characters of the novel as narcissi emerges from sketch-book journaling when it was re-discovered that one character wore a yellow suit, like Wordsworth's daffodils beside the lake, and another character took cover beneath the trees.

### The writer's sensibility as research instrument and as data

This interrogation of the literary texts associated with place, looking for colour, geography and connections to other literary texts begin to affect the sensibility of the researcher. Detecting where a novelist or poet has used language or place description to affect their readers and journaling those moments and linguistic methods creates a practised awareness that will make the detection more productive as the process continues. The researcher begins to see plays of memory both within the text under study and as memory references out to historical events that the literary authors share with their readership. If these are journaled, then the whole journal can become not just the source book from which to write the final travel story but also a collection of data that will shed light on the changing sensibility of the researcher. This data analysis may well be one of the research aims if this travel writing project forms part of a PhD or research Masters.

The overarching aim for the travel writer is to treat their sensibility as a research instrument being made more sensitive to more of the assemblage of the destination through reading and then in the field. How am I affected by this walk back down towards the quayside? Sensors of the images, sounds, the change in the gravity on the downhill slope, the evening breeze turning to blow in off the lake, the aroma of cooking from the restaurants as they open, are constantly recalibrated by new ways of speaking about the feelings experienced that draw on new metaphors and similes. The readers want to have trust in this sensibility that it is living and becoming more entangled in the place they want to know through this narrative and subsequent visit.

Thorough journaling and keeping of field notes will provide the data for the researcher to see any developments and gaps in their sensing of emotional and cultural space. For example, what notes did I take on the architecture of the façade of the old customs house as it came into view? How can I draw on my memory of other authors to make that more affective for my readers?

## Knowledge management tools in researching and travel writing projects

As a worked example of a Web 2.0 platform for travel writers, this section expands on the journaling designs begun in Chapter 4 and shows how these can be applied in the readily available software application from Microsoft, called OneNote. At the core of journaling is the ease of making a note and then returning to that note via rapid searches, so a system of editing and coding with meta-data is needed right from the outset of maintaining accessible journaling. In this section, the established principle of zettelkasten is developed within a software application that is accessible on laptop, tablet and smart phone, so that the digitised and indexed journal is always with the researcher-writer.

During the research for this book, Google Documents and Drive were first explored as a shared platform for DMO and writer (Mansfield & Potočnik Topler, 2021). However, a template was also designed and refined within MS OneNote to initially offer a personal knowledge management system for the researcher. During the research, experimentation unlocked the cooperative aspect of Web 2.0 and cloud-stored files, and this is explained here (Table 5.6). First came the OneNote page design, which can be set within the software application to be a default template for whole projects.

The pages are kept in only three distinct OneNote sections so that the overview of the journal is not too complex to see at a glance. These three sections are also the structure of the three-step process proposed by this book for commissioned work or research and use the colour scheme of a yellow-ochre for deep-mapping, green for fieldwork and purple for final pieces of literary travel writing (see Table 5.7).

**TABLE 5.6** MS OneNote Zettelkasten page template for travel journaling

| | |
|---|---|
|  | *zKeywords z…* |
|   Enquiry, quote or theme | Catalyst for this OneNote page: |
|   Affirmation journaling | |
|   Emerge | Emerging questions to carry forward |
|   Links | To other OneNote pages and to URLs |
|   References | |
| Dialogue | |

Designed during the research for this book (Mansfield & Potočnik Topler, 2022).

**TABLE 5.7** The commissioned travel writer's project process shown within its MS OneNote format

| Name of Step | Source Documents | Dialogue with | Outputs |
|---|---|---|---|
| 1. Deep-mapping and Route Design | Maps, novels, diaries, biographies, reports on land use change, scientific articles | Locals, stakeholders, academic researchers, for example, geographers, tourism and heritage specialists | Journaling. Sketch maps of planned routes for walks Trust-building correspondence |
| 2. Fieldwork | Menus, leaflets, posters, local newspaper, receipts, and the live environment of the field experience | If possible live dialogue with remote colleague, mentor or teacher, via email or on this platform Interviews on the spot | Field notes Theme, sub-theme and twill elements for travel story |
| 3. Recounting the travel story | Own field notes. Additional desk research to complete details and fact check | An author-editor if available to read drafts. Blogging extracts for feedback from trusted stakeholders | Finished travel story for publication in various formats or for digital delivery to commissioning stakeholder |

Designed during the research for this book (Mansfield & Potočnik Topler, 2022).

The Zettelkasten page, shown earlier in Table 5.6, with its icons and table structure is set as the default template for every page in all three sections; this is to speed up the creation of pages whenever an idea or quotation stimulates a moment of journaling. An overview of all the pages of notes can be displayed by selecting: Review, Linked Notes. The working writer, as the project matures, needs sketchbook-style views of all the pages in OneNote in order to make further connections and to be reminded of ideas from earlier in the work, so this review feature is useful. Also valuable for overviewing the whole set of journal entries is the search option, which can be refined with the zKeywords prefix to find only those journal pages coded with the zed.

# 6

# CASE STUDIES IN COOPERATIVE TRAVEL WRITING

## Introduction to cooperative travel writing

These case studies are presented to show the working between literary travel writers and their author-editor. The extracts are from live research projects that have been operationalised in the field to explore and then demonstrate the trust-building between two writers. The use of dialogue and the change in questioning are illustrated as a model for future research work in this method of cooperative working. At a very basic level, the author-editor is a copy-taster, pointing out a name or a position in the field that, although obvious to the travel writer is obscure or confusing to the final reader. At the next level of contribution, the author-editor uses more specialist linguistic vocabulary to encourage the writer to take more risks with the story. This may be to include a new theme or to edit the language to a more poetic or emotive register. Finally, often the most difficult is to see themes or ideas that may not be effective and to find ways of improving them or suggest removal.

The relationship between literary travel writer and author-editor is key to a successful writing process. It is important for the travel writer to feel that the author-editor is well intentioned and can trust them and that the writer is not offended by any comments and suggestions. It is desirable to have an exchange of opinions and discussions on different views and perceptions, however. On the other hand, it is equally important that the author-editor acts encouragingly and, at the same time, maintains as objectively critical an attitude as possible, despite any prior sympathy or antipathy of the writer. It is also very important that the author-editor tries to give instructions on how to improve the text.

DOI: 10.4324/9781003178781-7

## Practice-based field work in Slovenia and Montenegro

## Case study 1

### The *Both* project in Slovenia

These case studies track and explain travel writing projects. The first one is from early in the cooperative writing and explores Ljubljana and the route to Koroška in Slovenia. It is hoped that readers will first check the location of these places on an online mapping system. The project called, *Both*, was both creation and experiment and ran through July and August 2019. A literary figure was chosen as the catalyst for the travel writing quest. The Slovenian author, Prežihov Voranc is commemorated in Podgora, with his former cottage now open to the public. Search for Preški Vrh 13, 2390 Ravne na Koroškem, Slovenia, in an online map system. Note on referencing: participants' first names are used to reference field notes and place-writing, as below:

> I sat down on the terrace and rested my eyes for a while on the stunning steep green hills.
>
> *(Jasna 19 July 2019)*

The work began with text that can best be described as travelogue. This type of writing uses the eternal present tense, thus setting the tourist destination in a permanent state, where people, places and practices stand still in time. Consider this from near the opening of the writing, as an example:

> Of course, there is also an airport in Ljubljana. To reach Podgora, you will need about 2hrs by car. If you have the chance, visit Ljubljana and its Old City Centre with Plečnik's Triple Bridge and Prešeren's Square on the way, and perhaps take a boat on the Ljubljanica River. The capital abounds in magnificent architecture, pleasant cafes, restaurants, and cultural events.
>
> *(Jasna 12 July 2019)*

The nature of the *Both* project created an interested reader for the writing produced; this accounts for the direct second person address of, you, as in 'you have the chance', calling into the journey the cooperating researcher. The intent was that the addressed researcher would visit, following the lines of the journey. As a method, it produces a direct second person address which creates a written text that engages new, later readers with this 'you'. A kind of a dialogue is formed.

Dialogue has been a method of teaching since Ancient Greece when the philosophers, such as Aristotle, Plato and many others exercised and developed it, but it is also a useful research tool used for 'clarifying positions and creating new understandings' (Burbules, 1993; MacInnis & Portelli, 2002, 33). But what is a dialogue? Is it a conversation about a certain topic? MacInnis and Portelli (2002, 34) state that there is a difference between the two as 'conversation aims at equilibrium while dialogue encourages disequilibrium in order to advance given arguments', dialogue is also more focused. Thus, dialogue has contributed significantly also in this research on travel writing. Authors not only discussed their work, but also commented on each other's theoretical and practical contributions. Dialogue, usually in the form of scientific debates, enabled the two authors to identify what needs to be clarified, more explicitly clarified or further explored. They asked each other questions, gave each other instructions, had debates and commented on the work done. Due to the distance between the authors (one has been working in the United Kingdom, and the other one in EU, Slovenia), the dialogues mostly took place via e-mail or MS TEAMS and ONENOTE, and through commenting on the text in shared files. Dialogue facilitated the processes of researching and writing for this monograph. Researching through dialogue is more successful and rewarding than solitary research, as Shor and Freire (1987, 98–99) point out:

> Dialogue is a moment where humans meet to reflect on their reality as they make and remake it. Moreover, to the extent that we are communicative beings who communicate to each other as we become more able to transform our reality …

It is very important to note that in dialogue 'each discovery of a piece of evidence helps the participants to see what additional evidence might be necessary or what additional questions might need to be explored' (MacInnis & Portelli, 2002, 35).

However, by 19 July 2019, a journey for fieldwork had been planned and carried through, with a shift in tense for the created texts. The cooperative nature of the project is shown using transcriptions, below, of the dialogue questions, each followed by the section of travel writing that prompted that dialogue question and then a section of new journaling by the main travel writer as she prepares further resources for a later write-up o the final travel story.

## *Peer dialogue 1 – Narrator's connection with history and place*

In this second paragraph from your field notes, Jasna, I can see that the quest is to reach Podgora. However, readers still have no connection with the narrator. Modiano uses historical events to place his narrative in time and place and simultaneously connects historical events with the narrator's own life. My first dialogue question, then, is: Do you have a clear childhood or school memory of Podgora, and does that connect with a historical event in Slovenia's history, for example, summer 1991 when the Brioni agreement was signed, or spring 2004 as Slovenia prepared for accession to the EU?

> To reach Ravne na Koroškem and from there Podgora, you can take the regional roads Ljubljana – Domžale-Trojane (you must stop here for the best traditional Trojane doughnuts) – Vransko (restaurant Grof is a great choice if you need a decent meal or just a quick bite on the way) – Velenje (in the past the city was called Titovo Velenje – Tito's Velenje, here you can stop to visit the Velenje castle, the Museum Of Coal Mining, every September the Pippi Longstocking Festival is organised here [...] (Jasna field notes 2019)

### Journaled response to dialogue 1

My primary school was named after Lovro Kuhar – Prežihov Voranc. Consequently, the author was discussed a lot during my primary school years. I still remember some parts from author's works by heart ... And once a year – usually during the spring hiking trip, we visited Podgora and his cottage – the place called Prežihovina – this could be translated as the estate of the Prežih family. For everyone going to the Lovro Kuhar – Prežihov Voranc Primary School, this is a special place, and for others a tourist attraction since Prežih is a canonised author. There of course remains the question of how to attract international tourists to this place. One of the solutions for attracting foreign visitors and tourists is to find references to foreign places in Prežih's life and works, and to use these references as elements of storytelling related to Prežih and Prežihovina. It is well known fact that when Prežih was hiding from regimes because of his activist and political work, he was travelling across Europe, spending some time in Austria, Italy, Germany, the Czech Republic, Greece, Bulgaria, Romania, France, Russia, Norway, etc. These places could be used as referential points for attracting tourists and employed as tools for attracting tourists and visitors in travel writing. Travel writing texts can further on be offered for reading and engaging tourists in travel writings. Prežih's collection of short stories titled Solzice (Teardrops), along with his works Samorastniki (Self-Sown), Požganica, Jamnica and Doberdob, which comprise a foundation for creating new stories, notes and travelogues, can be utilised as tools for attracting wider audiences and international visitors. The next step is the tourists' guided and self-directed movement to the places mentioned in the texts.
...

### Peer dialogue 2 – Narrator's memory activated on Slovenian culture

How is cookery taught in Slovenia in 2019? Do the teachers separate boys and girls to teach them separate subjects, woodwork and domestic science?

Or has there been a change so that everyone or no one learns to cook? Can you remember any cookery classes?

> Still, it is good to stop in Trojane, a small village that represents the invisible border between the Ljubljana and the Styria (Štajerska) regions. Most people stop here for big delicious doughnuts – 'trojanski krofi', but you can also grab a meal in one of the oldest Slovenian restaurants with huge parking space and a nice terrace with hills view. (Jasna field notes 2019)

*Journaled response to dialogue 2*

Gastronomy is a significant part of cultural heritage of Slovenia and consequently an essential part of the contemporary tourism offer. With various reality shows, cookery is really popular also with younger generation and primary school children. I remember my cookery classes in my primary schools – we were cooking and discussing cookery once a week for two hours in the sixth, seventh and eighth grades. And since our teacher was among the most popular teachers at school with a very positive attitude towards life and everyday problems (her surname was Trafela), these classes were pure joy.

  …

*Peer dialogue 3 – Narrator's memory activated on Slovenian Social Space in Media*

Media from the public sphere reactivates the past, and stimulates nostalgia whilst giving readers and the writer strong reference points in recent historical time. The Slovenian literary journal, *Sodobnost*, provides a connection with the literary author being researched during this field inquiry. The journal was still published in 2019 during the field work. Were any articles of interest to the theme of nostalgia in the June issue?

Can you find an old copy of a newspaper from July 2019? For example, *Delo*, or even better for its literary connections, *Sodobnost*. What was in the news then? The other people around you might be reading newspapers. Do *you have any memories of your parents or grandparents reading a newspaper regularly? Which one?*

> 'Dober dan. Kaj boste?' 'Dober dan. Štiri krofe, prosim', 'Z marmelado?' 'Z marmelado, prosim', 'Izvolite'. 'Hvala'. Many people, colourful Slovene dialects mixed with some German. The service was speedy. I went for some cold juice and doughnuts for later. 12 euros and 30 cents altogether. Receipt dated 19 July 2019 (Jasna field notes 2019)

*Journaled response to dialogue 3*

Yes, Saturday's *Delo* with its substantial supplement called *Sobotna priloga* was traditionally obligatory reading in our family. The supplement was a synonym for quality journalism. Unfortunately, the glory days of this newspaper are over, as the viability of printed media has drastically changed. Another widely read newspaper in Koroška, the region of Prežih, is *Večer* (večer means evening), which is probably the most popular newspaper among the readers in Koroška; they wrote about Prežih's museum in September of 2019. This is the title of the article written by a correspondent from Koroška (Večer's headquarters are in Maribor), Andreja Čibron Kodrin, the wife of my high school literature teacher Miran Kodrin at the Grammar School in Ravne na Koroškem:

'Z drugačnim pristopom in novimi vsebinami do bolj zanimivih spominskih sob'

'With a different approach and new content to more interesting museum rooms'

...

## Peer dialogue 4 – Narrator's memory activated to connect Prežihov Voranc in time and place

How old was Prežihov Voranc, Lovro Kuhar when this restaurant opened? How does that compare with your daughter's age? How old was Prežihov Voranc when his book *Vodnjak* (The Self-Sown) was published? How does that compare with your career? How far away from the Grof restaurant did Prežihov Voranc live?

> But the other day when I was with my family, we made a dinner break at the Grof Restaurant – another one with an enviable tradition – since 1901. A hundred years ago this was a stop for coachmen, today some exhibited objects, photographs and furniture hide and reveal stories about the way of living of our great grandparents. On arrival, the smell was tempting and the atmosphere great. And the parking areas were full, which is always a good sign of the service quality. My daughter immediately started asking about what various exhibited objects and tools were used for in the past (Jasna field notes 2019)

*Journaled response to dialogue 4*

In 1901, Prežih who lived around 75 km from Vransko was 8 years old – my daughter's age. But his childhood at the beginning of the twentieth century is probably incomparable to the childhood of today's kids. My grandpa told me stories of his early childhood – he is nearing his 90s – and he had poor diet – potato soup every day, sometimes twice a day, and as a very young boy, he had to earn for this poor food as a shepherd at a nearby farm. His mother had four kids and because she could not feed them all, my grandfather as the oldest son in the family had to go away to earn for his own bread – literally.

### Peer dialogue 5 – Narrator asked to find and insert Free Indirect Discourse as a social comment within the travel story

Free Indirect Discourse was introduced in Chapter 2 as part of the concept of heteroglossia. This allows the travel writer to insert text which is not attributed to her as a narrator, but taken either from a voice or text she encounters or an unattributed phrase that makes a moralising or ironic comment on the practices observed in this tourism space.

24 July 2019

Yesterday, I read that in 1943, Prežih was imprisoned by Gestapo in Begunje na Gorenjskem – as a former high representative of the Kominterna. It was hot as hell, but I decided to drive to Gorenjska anyway, who knows – perhaps I find something related to the famous author.

Oh my goodness, it's five in the afternoon, but my car says that it's 43 degrees outside, ok, after a while it turns out it's only 37.

When I am approaching Gorenjska, I see Julian Alps in the distance. It's not far to Begunje na Gorenjskem – I took the highway, so I needed only about 20 minutes by car.

A big accordion in the middle of the roundabout reminds visitors of the Avsenik Museum. Slavko and Vilko Avsenik are the authors of Golica (Trompeten-Echo in German), the most often played instrumental melody in the world. The melody belongs to the genre of folk music. I leave the roundabout and follow the sign Begunje na Gorenjskem. By the way, Begunje is famous for the best skiing designers in Slovenia – Elan, championships have been won with »elanke«. No sign of any Gestapo prisons from the WWII. No, I was wrong. I saw a sign for the memorial museum. A local told me that the former Gestapo prisons were turned into a museum. More than 11,000 people were imprisoned here, some were shot and some later moved to concentration camps. Also, Prežih was moved to a concentration camp. From Begunje, he was transferred to Berlin, where he was offered the position of a president of Slavic provinces under the German protection. Because he declined the offer, he was put to Sachsenhausen. (Jasna field notes 2019)

### Journaled response to dialogue 5

The region of the Julian Alps is magnificent for hiking and mountaineering enthusiasts. In *The Guardian*, Chis Zacharia (Zacharia, 2021) mentions the Juliana Trail, 'a route covering 270 km, split into 16 stages on a circuit around Triglav'. In his travel writing piece, he says:

> I'm in Slovenia but I'm not anywhere easily found on a map. Here, at the far east of the Julian Alps, is a land of water and woods. Half the country is forested, but the part we're walking through is especially remote. Triglav national park is one of Europe's oldest. In the middle sits the eponymous Triglav, Slovenia's tallest mountain, spiritual summit of the nation's fables and myths.
>
> *(Zacharia, 2021)*

---

## Case study 2

### *Memories from Montenegro*

Note on format: The texts scanned using Google Lens are from the field notes and are presented in short sections, research discussion and evaluation of the scanning is given in plain text below each section:

18th July 2021, Tivat, Montenegro
This morning I read about terrible storms across Europe, [...]
It is 12:55 and I sit on beach [...] near the Tivat airport
[...] we are all sitting in the Orange Beach Bar – around 10 guests.

The researcher-writer in the field clearly sets a date and time for the author's editor and later for her readers to pinpoint her location. In tourism development and city branding, the plateaus, those places chosen to write from, provide mapping points for later visitors who want to follow in the footsteps of the literary travel writer and share their experiences. The Orange Beach Bar can be found by the readers and the cooperating editor by using Google Maps. The beach looks out onto the Bay of Kotor, locally called the Boka on the Adriatic. The travel writer faces a challenge here, especially writing for another language group, for example, Anglophone, because if she starts to include these place names it might overload the readers. However, if a hexis or dialogic map is being created, then some method of drawing with words is needed to build deictic points of reference in the readers' minds (Mansfield, 2020, 6–8).

I ordered a Cockta – a drink similar to Coca Cola, but so it is believed, more healthy and invented in Slovenia in 1953. [...] Have to run ... the rain is here

The scanner tool transcribed this text as 'I ordered a cell the and nobody around quests. – a drink similar to Cockta – Coca Cola, but healthy so it is and invented 1953', so a lot of post-editing required to render it as the original field researcher had meant it to be. In other tests, at the laboratory, it was found that lined paper in a hardback notebook provided the necessary staves to guide the scanner and help it maintain straight text lines during reading.

The theoretical integration going on in this section, though, is the discovery of local foods and drink, often called ethnobotany (Mansfield & Potočnik Topler, 2021). Explaining local foods helps the local producers and also guides the visitors to consume products that require less food miles to travel to the hospitality service outlet. If a story can be added to the food product, then some of the local culture can be shared with tourists, too, enhancing the value and satisfaction of learning.

20th July 2021 Tuesday, partly sunny, partly raining. I'm sitting on a ferry to Kamenari. My today's goal is […] to see the house of Ivo Andrić, the Nobel Prize winner for literature in Herceg Novi. The house has been turned into a museum.

The field researcher-writer continues her route mapping and introduces movement; one of the method points, labelled D4 (Mansfield, 2020, 8), in travel writing for tourism development since it indicates to the subsequent readers that a journey is needed to move to the next plateau, in this case a ferry journey. The scanning software did not resolve the name of the destination port, and this could not be found by the cooperating editor using Google Maps, unfortunately. This is a useful lesson for follow-up projects. Herceg Novi, on the other hand, is a town that is easy to locate. It was the site a new castle built on the entrance to the Bay of Kotor, which provides the literary travel writer with an opportunity for some local geology, explained in design method D12 (Mansfield, 2020, 11–12).

The next point introduced here is a quest. Further, this quest has a literary tourism value, to discover the work of Nobel laureate, Ivo Andrić. In the two-way correspondence between the researcher-writer in the field and the author's-editor back in the laboratory, this quest proved an exciting development. After reading the field notes describing the museum in Andrić's house, the journaling of the editor suggested this process of inquiry, drawn from Sebald's, *Campo Santo*:

Activate in the field – *Journal of Author's Editor*
p. 44 of *Campo Santo*: back in his hotel room in Piana, Corsica, Sebald reads Flaubert's story of St Julian in the Pléiade edition. – J. in fieldwork the literary travel writer needs to do activities like this that *anarchive* the place, that is, unlock it, activate it with diffractive practices, following new posthumanist practices.

For example, 'in the drawer of the bedside table taken from the archive and played to diffract the light from the window'.

The proposal here is to include a reading from the novel that the researcher carries with her in the field, for example, *The Bridge on the Drina* (March 1945) by Ivo Andrić. However, this needed more desk research to discover either a novel that Andrić set or wrote in Njegoševa Street or, if this is not possible, then to take a major theme from Andrić's writing oeuvre to work with in the literary travel writing. One example plateau where this post-trip research could be re-inserted when converting the field notes to a publishable story is here:

20th July 2021 [...] [Andrić], an intellectual writer and a diplomat, won his Nobel Prize in 1961. I wonder what was going on that year in international politics.

Later, desk research reveals that in the run-up to 1961, Andrić published a short story *Panorama* (1958) with the publishing house *Prosveta*, in Belgrade. Then, in 1960, *Priča o vezirovom slonu, i druge pripovetke*, a collection of short fiction, again with a Belgrade publisher. His theme is that by understanding the history of other cultures people are drawn together in peace.

A more immediate method of activating in the field is to include conversation or stories from others. In the café-bar scene, for example, the stories of the other guests could be integrated in the manner exemplified by Rötig (2018); in her travel book, *Cargo*, Rötig waits in a bistro near the railway station in Le Havre, and from that plateau on her hexis recounts her testimony of the storytelling of Élodie.

### Post-fieldwork dialogue questions and desk research on Andrić's life and work

*The Bridge on the Drina* (March 1945) by Ivo Andrić does begin with literary descriptions of the geography of place.

For the greater part of its course, the river Drina flows through narrow gorges between steep mountains or through deep ravines with precipitous banks. In a few places, only the river banks spread out to form valleys with level or rolling stretches of fertile land suitable for cultivation and settlement on both sides. Such a place exists here at Višegrad, where the Drina breaks out in a sudden curve from the deep and narrow ravine formed by the Butkovo rocks and the Uzavnik mountains. The curve which the Drina makes here is particularly sharp and the mountains on both sides are so steep and so close together that they look like a solid mass out of which the river flows directly as from a dark wall. Then, the mountains suddenly widen into an irregular amphitheatre whose widest extent is not more than about ten miles as the crow flies.

Here, where the Drina flows with the whole force of its green and foaming waters from the apparently closed mass of the dark steep mountains, stands a great clean-cut stone bridge with eleven wide sweeping arches. From this bridge spreads fanlike the whole rolling valley with the little oriental town of Višegrad and all its surroundings, with hamlets nestling in the folds of the hills, covered with meadows, pastures and plum-orchards and criss-crossed with walls and fences and dotted with shaws and occasional clumps of evergreens. Looked at from a distance through the broad arches of the white bridge, it seems as if one can see not only the green Drina, but all that fertile and cultivated countryside and the southern sky above.

Post-fieldwork dialogue questions posed by cooperating author-editor to the literary travel writer:

1 What effect do these descriptions create now that you have visited the region?
2 Do you think Andrić is interrogating and describing the places for any political or thematic effect?
3 What was Andrić's relationship with the town of Herceg Novi? How has that been interpreted in relation to his writings? Are any of his writings set in Herceg Novi?

## Place-writing experiment on Cherbourg as a new tourism space

### The contribution of Roland Barthes to writing studies

Roland Barthes (1915–1980) left an important legacy on critical approaches to literary style in his essays on writing practice and in his two works that are of interest to literary travel writers. These are his autobiography, which appeared in France in 1975 and was quickly translated by Richard Howard for the US market in 1977 (Barthes, 1988), and his travel book on Japan, and Tokyo in particular, *Empire of Signs* (1983). The travel book ostensibly examines the urban space of Tokyo from Barthes' point of view as a tourist and as a visiting scholar. It was published as a collectible artbook by Albert Skira in Geneva in 1970. Its readers are therefore offered the signs that this is travel writing to be savoured and enjoyed, but as the text proceeds, the work oscillates between an ethnological discourse and a narrator concerned, almost anxious about language. Barthes, it appears, chooses this genre: the travel story, to examine his language theories in himself and his own shifting identity as a writer. In particular, he makes it clear that travel narrative anyway is a kind of speech that offers the writer the opportunity to move between the narration of history, both personal and social, and the exposition of an ethnological study.

Barthes, although holding a university post at the College of France, published his books without the critical apparatus of references normally expected in academic work. This mode of working gives a writing that is playful and full of challenges or at least a writing style where the reader must carefully seek the interpretation. Barthes writes in this way deliberately since his colleagues in the late 1960s were analysing the production of meaning by writing. In fact, Julia Kristeva and Michel Foucault proposed at this time in the 1970s that language determines the meaning of speech or text by controlling its users (Kristeva, 1980). Foucault asks questions about scientific texts, for example, ethnographies (Foucault, 1970); he wonders if the texts dominate their subjects, but he also proposes that language is the site where we can imagine the impossible: where two subjects can be placed immediately adjacent to each other, whereas in the place itself, they could never be seen together. He proposes that this is where readers and writers juxtapose objects in the non-place of language (Foucault, 1966). Barthes takes on this idea because his writing is precisely the act which unites in the same work, those subjects which could not exist together in a single flat space (Barthes, 1988). This begins to build for Barthes two of the key ideas of his travel book, that of utopia, the ideal place, and the void, or the empty centre.

The chapter on the traditional, historical centre of Tokyo city, the forbidden palace of the emperor, introduces the empty centre which Barthes contrasts with the full city centres of the west (Barthes, 1983, 30–42). However, after a short personal recollection of how Tokyo's city addresses are communicated between colleagues, he explores the railway station and finds a centre which is full of activity and identity but still empty spiritually. At a named district rail station, Ueno, Barthes brings us close to his experience as a visitor in this vast underground city, which he says, 'finally fulfils the novelistic essence of *the lower depths*' (Barthes, 1983, 39). A glance at an online map of Tokyo shows that Ueno is the station that serves the tourism destinations of the museums, zoo and park of the city. No mention of these appears in Barthes. Following his trajectory across Tokyo would be something for the specialist in writing, and someone interested in what stimulated Barthes' essay-like fragments.

A few years after the travel book commission, he embarks on his autobiography (Barthes, 1988) and turns again to the fragment as the building block for the whole book. Rather than essays, these pieces typically spend a paragraph contemplating a single idea. And, in his challenge to narrative, he uses alphabetical order to give side-headings to each subject he treats in each fragment; this is also a formal challenge to the chronology of the life being recounted. Like a literary travel writer, he lets himself be influenced by his fellow travellers, for example, under the letter 'I' '*The subtle instrument*' (Barthes, 1988, 107) Brecht's concept of the tiny, fragile instrument is introduced to tell his readers that he feels that the world needs putting back together, but that the operation will be delicate.

A travel story could follow this same alphabetical ordering, yet still retain the route of the writer; the letters would sharpen the observation and creativity of the researcher-writer to find a named object or concept that matched the letter at that point in the streetscape. One photograph in Barthes' autobiography is simply documented 'Cherbourg, 1916' (Barthes, 1988, 21). It is a black and white photograph of Henriette Binger seated, holding Barthes, who is just a few weeks old. The frame is oval or elliptical suggesting it was produced by a professional photography studio. Binger lived in Cherbourg at the time of her son's birth, 12 November 1915. Her husband was often away as sea, and was killed in World War I. Given Barthes' importance to writing studies, finding his birthplace is of value to literary tourists. Henriette Binger's address in 1915 was 107 rue de la Bucaille, and since this street name still shows on maps, it could be the destination for a travel writer's walking route through the port town of Cherbourg, and thus the recounted route could become the armature on which to build the literary travel writing piece.

### Plotting a route in the unknown town

During research, travel writers will frequently find themselves arriving in a town unknown to them, even with guide books and online mapping systems with street-photographs, the far-off streets of their route are mysterious and still surprising when encountered physically. Indeed, in this example in Cherbourg, the mapping system showed that there is a house numbered 107, on the street under investigation, when in fieldwork, it could not be found. For the Cherbourg travel writing case study, then, the birthplace of Roland Barthes was chosen as the furthest point to walk to and document, since this was considered to hold most value for later visitors. A tourism management aim was to render the city of Cherbourg a worthwhile stopover destination for visitors arriving by ferry or train, who would normally travel onwards directly without considering the urban space of leisure value.

A preliminary route is planned and written out in the hardback field notebook so that a version is available to add to and refer to during walk. Route is planned using online mapping software. Transcription shown below:

**Transcription from Field Journal, Wednesday 11th May 2022**

15 minutes walk. Take the Quai Caligny, turn in left off the quayside in front of Ambassadeur hotel, west into
    rue du Port, then after 4 buildings [Note: Detour first to 50 Rue Grande Rue to see bookshop, Au Petit Bouquiniste, then back] left again into rue des Fossés. Long walk then right into
    rue du Commerce then left into rue au Blé, a square with trees and a green circular building turn right into

> Pl. de la Fontaine which becomes rue Christine very long walk, cross
> major intersection into rue de l'Ancien Hôtel Dieu. then take
> right fork into
> rue de la Bucaille very long walk to 107.

The following day the field note book was used to record the journey, in the field as close in time and place as possible to the position being observed and described:

### *Transcription from Field Journal*, **Thursday 12 May 2022. 07h 55 AM.**

1.2 Kms route from Google Maps 07h55 am Thurs 12th May 2022 Cherbourg. Good French coffee at last for breakfast. Warm, with small high clouds in blue sky In the field! – turning left into rue du Port now, little cooler out of the sun. Rue des Fossés is narrow but just going to look for bookshop in next street in Rue Grande Rue. The building at the end of this street might be the blue schist; it is an immobilier. Found the bouquiniste, closed 'til juin. Now back out to Fossés.

Rue du commerce is all shops, in good condition. Trade is good, 2 clothes shops closed down, small independents. The green circular building in Place de la Fontaine is the fountain!

8h19 found rue de la Bucaille. No street name so asked someone. Very helpful smiling. Quiet road, very few moving cars but pedestrians beginning to appear. At number 4 *bis* the house of Dr Prosper Payerne who invented precursor to the submarine, died 1886.

Huge block of a building opposite entrance to rue de la Comédie. What was it? Warehouse, convent, prison? Row of 14 windows high up at the top.

Rue de la Bucaille stops at 63 and 68. at junction of rue Asselin and becomes. rue Pierre de Courbertin but I'll continue for a few hundred metres …

8h45 am. No, de la Bucaille ends there. Although Google Maps places 107 in the rear of this pair of quite elaborate stone houses with balconies on 1st floor and pair of mansardes. [NOTE ADDED: See the timings from start to arrival is 55 minutes].

On the way back, fascinated by the entrance a long arched tunnel beneath a public building dated 1871 LXXJ is clear but MDCCC damaged.

Liais Gardens, first on your right, the garden of fragrances.

Trying to change route on way back at junction of rue Christine to walk down long straight rue E Liais to blue schistes bookshop. It was closed until 10 am but provided a reference point for return route. Just walk down rue Gambetta off this circular Square to peek into Impasse Couppey at the schist buildings then cross to go right.

went into rue Victor Grignard in attempt to arrive at library building. Do not be lured by the green hill but tum quickdy left inlo Pass. de l'Alma. It was well worth it because here I discovered la Résidence Roland BARTHES, Passage Loysel. Out of Pass. de l'Alma inlo rue Jules Dufresne.

Crossed a huge post-war street, Schumann, I think. 1970s feel to architecture.

I came out onto a street market where 4 stalls were roasting and smoking pork. 9h45 end of walk, back in the square with the statues.

~

Back in the hotel for an hour until 11 am. Large grey clouds now starting to cover the sky. The best of the day is over, and I caught it all. Especially exciting was the chance discovery of La Résidence Roland BARTHES. I was drawn there, first by trying to add 2 bookshops into the return route and then realising I could end in the library square where the new bronze statues are placed. I stayed open, and let the field of the street draw me there Saw, too, that the inland-most windows of Hotel Mercure will have Line of sight into the library square, to see Barthes statue.

From these field-notes and further reading, the following literary travel writing is constructed. One major theme taken from travel book on Tokyo city by Roland Barthes (Barthes, 1983) is the contrast between a busy urban space and an enigmatic empty centre. Readers who know or who are reading Barthes should start to detect this theme emerging in the travel story.

## The plateaus of the travel story for public readership

### L'Ambassadeur – The Ambassador

I had noted L'Ambassadeur hotel on Wednesday evening because it looked out across the Avant-Port of Cherbourg's harbours. It also acted as a turning point to lead walkers off the Quai Caligny and deep into the streets of the old town. Often, no amount of online photography can convey the real aspect of a building. The Ambassador was like this and, when I saw that its street number was 22, then it became for me, the starting point for the route to venture inland, across the town in search of the former home Henriette Binger.

### Bouquiniste – Second-hand book-dealer

By 7:55am on a warm Thursday morning, 12th May 2022 1 was already on the Quai Caligny. Cirrocumulus clouds in a blue sky suggested I would need no umbrella. Henriette Binger was only 22 in 1915, when she came to live for a short time here in Cherbourg. She had married a naval officer in February 1914 but when war broke out towards the end of their first summer together, it meant that he would be needed in the Channel and North Sea.

They travelled to Cherbourg for him to report for duty. A left turn beside the hotel took me into the rue du Port. Immediately, it was a little cooler out of the sunshine. I walked quickly to the opening which was the rue des Fossés; it was narrow, as I expected from its name, which often meant dry ditches in medieval France. But I wanted to go to look for the second-hand bookshop in the next street, in rue Grande Rue, which used to be enclosed by the château, demolished at the end of the 1600s. The enclosing wall of the château no longer existed. The medieval town centre had gone, but the local building material was presented magnificently by the shop at the end of this street, 49 rue Grand-Rue. A canopy of blue stone tiles sheltered the door from rain, and the outline of a narrow arched window on the first floor made me think of the lost château. I imagined the occupant looking down on the busy street scene. Could they see as far as the port?

I found the little bouquiniste at number 50, but it was closed until June. Closed bookshops have a special intrigue. They promise so much, like book covers, but remain essentially empty for the eager reader.

I turned back and retraced my steps to the opening that was the street of the dry ditches.

### Notes on Cherbourg work

By adding two dietic markers, a plain statement of fact about Binger's life can be improved to place the reader with the narrator in the urban space: 'when she lived in Cherbourg' was changed to 'when she came to live for a short time here in Cherbourg'. The adverb, here, and directional marker in the verb, came, rather than went, or stayed, are difficult to add once the writer has left the scene of the fieldwork, for example, if writing up the travel story back in the hotel or even on the ferry back to home. This is because the writer's own centred point of view has changed. So much so, that an inadvertent use of a directional marker can confuse the reader as to the whereabouts of their narrator during the story.

### Dialogue questions from author-editor

*The travel writer's notes offer a very useful insight into the process of travel writing. I would like to ask the author also the answers to the following questions: a) Did you have enough information recorded in your field notes?, and b) What kind of notes can be created during the field trip? By that, I mean, what were the difficulties of writing on the move, outdoors? What could you do to capture more points out there next time?*

*When I was copy tasting, I thought that the number 22 seems significant to the author but not to the readers. Can you include a little help to draw out the importance you feel for 22? Naming the clouds adds an interesting fact to the text and makes the reader start asking about the writer's identity or character. Who is the writer? Someone who is interested in clouds? A geographer who knows the names of the clouds? Someone used to observe the weather. How is the weather related to the mystery character of Henriette Binger? I think it needs a link sooner here or leave out the detail of the clouds. Are you hinting at the link between umbrellas and Cherbourg from the film? If so, your readers need a little more help.*

*Adjectives (turning, old, online, real, former) are used in just the right measure to diversify and embellish the text.*

*How do you plan to maintain that theme started by the closed bookshop or empty book? Is this going to connect with the concept from Roland Barthes of the city with an empty centre? Are you sure you mean that books are empty or do you mean that when you can only see the cover, then the readers are unfulfilled?*

Here are live examples of the author-editor encouraging the literary travel writer to develop more from their first 3 plateaus (Ambassador, Bouquiniste, Capons).

*J says: This part in Capons, below – imagining the past and historical events – is effective: 'In this one view I could see how this had been a space of trade', but the inclusion of the French word, solier, is an overload, because the readers are already dealing with the word 'capons', which will be new to most readers; and the French street name, Fossées. The flashback to the medieval street offers you the opportunity to make the scene seem busy; you need this busy mood if you want to contrast later with an empty centre. You say 'a space of trade' which sounds too formal as a register for a flashback story, but this is the point where you could create the busy feeling.*

## Capons

I had only walked past the first building in rue des Fossées when I saw a flight of steep steps running up beside a tiny bar, that used to be called Le Solier. In this one view I could see how this had been a space of trade in the Middle Ages. I could imagine how vendors used to sell capons, from these terraced slopes to customers walking on the dry earthen floor, the *solier*, where I was standing now.

Further along the pedestrianised street, when I touched the greenish-blue lamina of the building stone it seemed crumbly and dusty. It was difficult to believe that it had survived so long. Then, at the further end of the street at number 3, a narrow doorway with two keystones in a cream-coloured rock, formed a slightly fallen ogee arch, and I was reassured again that this was a medieval place.

*J says: The word 'ogee' is too technical. Do you plan to use it somehow later? If not, then swap this out for a more imagistic phrase that readers can picture readily. Or do you think it conveys your idea if you simply remove that word? 'a slightly fallen arch'.*

# 7
# EVALUATING WRITING FOR QUALITY AND VALUE IN MENTORING

## Evaluating quality in writing

This chapter works through the interaction between the travel writer and the author's editor to show how the critical and creative relationship is built. Trust is maintained by attempting to sensitively guide the text to address its original aims through effective feedback and dialogue.

First, though, it is essential to establish evaluation criteria that may be used to assess a literary travel writer's work, either for themselves or in the context of a taught course with a marker or an external examiner:

### Evaluation criteria

1 Ability to identify an interesting and definable place and specific points along a route, the hexis (Mansfield et al., 2021).
2 Ability to develop a creative response in relation to the place(s), which involves emotional and ethical awareness.
3 Ability to develop a critical attitude towards tourism spaces, which tests existing boundaries.
4 Ability to add critical readings interpreted for a non-specialist readership from: academic journals, subject textbooks, newspapers, signs, novels, other travel writing, poems, specialised reports from geology, economic development, sustainability and social issues.
5 Style of writing, including register and grammar.
6 Degree to which the place-writing vignettes communicate successfully with audiences.
7 Development of work from an initial concept to final resolution, with documentary evidence of this, for example, sketchbook, journal, clippings portfolio, OneNote zettelkasten.

DOI: 10.4324/9781003178781-8

8  Originality of the final presentation.
9  A plan of publishing intentions.
10  In undergraduate assessment, the pass mark is usually 40% in UK higher education. Note, at postgraduate levels 7 and above this pass-fail threshold is 50%.

## A model design for a student feedback mark sheet

Teaching travel writing skills is both challenging and rewarding. According to Graham (2019, 288),

> effective writing instruction involves (a) Writing frequently for real and different purposes; (b) Supporting students as they write; (c) Teaching the needed writing skills, knowledge, and processes; (d) Creating a supportive and motivating writing environment; and (e) Connecting writing, reading, and learning.

At the Faculty of Tourism of the University of Maribor, travel writing has been employed for developing communication skills since the academic year 2019/2020. Examples below have been written by final year Master's students of the year 2021/2022. They display a variety of travel writing techniques and methods. These methods were taught and reinforced with further reading as the students prepared to go into the field to document their journeys. The example extracts below are from their post-fieldwork writing (Table 7.1).

Example (1) shows switching from the Past Simple to Present Simple Tense, and back to the Past Simple Tense:

1  Then we decided it is time for more history, so we started planning our trip to Pompeii. We were excited. I was excited. We couldn't wait to get off the train. Today, Pompeii is one the most famous Roman sites globally, but back in the days, Pompeii was just another Roman seaside resort until Mount Vesuvius erupted in 79 AD. Just a day later, thousands of people have been killed and the whole city was buried under a blanket of volcanic ash 25 meters deep, forgotten until it was unearthed by explorers 1,800 years later. I first heard of this buried city, preserved like a snapshot of Roman life frozen in time, in middle school history classes. Those books started my fascination with the Romans and put visiting Pompeii on my bucket list. (Lidija Kočevar, Rise From the Ashes, 2022)

Writing travel writing texts for an English readership in the Past Simple Tense is something students whose mother tongues, L1, are Slovenian and Croatian need to learn. Many of them would prefer writing in the Present Simple Tense. To mentor the students during the term, before final summative assessment, the lecturer acts as a mentor or author-editor to

**TABLE 7.1** Model design for summative student assessment mark sheet

| Criterion | Mark |
|---|---|
| **Content (maximum mark 30)** | /30 |
| I Outstanding work (26–30) Highly focused. Fully addresses brief with outstanding discussion of the main issues, with few errors. Demonstrates outstanding reading, industry and field knowledge. | |
| I Excellent work (22–25) Very focused on the assessment brief with some excellent discussion of the main issues, with few errors. Demonstrates extensive reading and knowledge of the field. | |
| II.i Good work (18–21) Addresses the assessment brief, with some good discussion of the main issues, using relevant facts. No significant errors. Demonstrates extended reading and field knowledge. | |
| II.ii Sound work (15–17) Addresses the brief using core information but some omissions. | |
| III Basic work (12–14) Addresses the assessment brief but contains a minimal amount of the required information. | |
| Fail – Limited work (9–11) Does not contain enough relevant information to address the brief or contains multiple errors. | |
| Fail – Unacceptable work (0–8) Clear fail that does not address the brief, with the inclusion of totally inadequate or irrelevant information and poor or inappropriate discussion lacking accuracy. | |
| **Use of relevant theory and literature including correct Referencing (maximum mark 30)** | /30 |
| I Outstanding work (26–30): Evidence of consulting an extensive range of valid sources of information, especially primary literature, and referencing this correctly. Outstanding application and integration of appropriate concepts and theories. | |
| I Excellent work (22–25): Evidence of consulting wide range of valid sources of information, especially primary literature. Excellent application and integration of appropriate concepts and theories. | |
| II (i) Good work (18–21): Evidence of consulting a good range of valid sources of information, especially primary literature. Good application and integration of appropriate concepts and theories. | |
| II (ii) Sound work (15–17): Evidence of consulting a limited range of literature. Some evidence of the application and integration of theory. | |
| III Basic work (12–14): Some reference to the literature evident. Statements made are not well–supported. | |
| Fail – Limited work (9–11): Some reference to the literature evident. Statements made are not supported as there is limited application and integration of the theory. | |
| Fail – Unacceptable work (0–8): None, limited or dated range of sources. No use of evidence to support arguments. No references. | |

/30

**Knowledge and understanding (maximum mark 30)**

I Outstanding work (26–30): Extremely well-constructed and logically presented and compelling argument throughout. Demonstrates an extensive understanding of topic within wider context. Full and highly critical evaluation, with analysis and arguments supported by evidence and examples. Some synthesis evident at higher levels.

I Excellent work (22–25): Extremely well-constructed and logically presented argument throughout. Demonstrates an excellent understanding of topic within wider context. Full critical evaluation with analysis and arguments supported by evidence and examples.

II (i) Good work (18–21): Well-constructed and logically presented argument. Demonstrates a good understanding of topic within wider context. Good critical evaluation with arguments mostly supported by evidence and examples.

II (ii) Sound work (15–17): A good assignment with some balancing of argument. Demonstrates sound knowledge and understanding of the key issues, theories and concepts but with little development. Evidence of some evaluation although limited with restricted use of evidence.

III Basic work (12–14): Enough understanding demonstrated but with minimal evaluation or evidence offered.

Fail – Limited work (9–11): Marginal understanding demonstrated that lacks evaluation of evidence. Fail – Unacceptable work (0–8): Limited understanding of the topic. Very poor analysis or none evident.

**Presentation, grammar and spelling (maximum mark 10)**

/10

I Outstanding work (8–10): Outstanding overall standard of presentation exhibiting a high standard of UK English and clarity of expression. Excellent layout and structure of material. I Excellent work (7): Excellent standard of presentation exhibiting a high standard of UK English and clarity of expression. Excellent layout and structure of material.

II (i) Very good work (6): High standard of presentation exhibiting a very good standard of English and clarity of expression. Very good layout and structure of material. Very good use of visual material and tables.

II (ii) Fairly good work (5): Adequate standard of presentation, using acceptable standards of English. There may be some lapses of expression and some attention to layout, structure and formatting may be needed.

III Adequate work (4): Weak presentation and structure with grammatical errors. Layout and structure may reduce impact and communication.

Fail – Limited work (3): Marginal standard of presentation. Poor use of English with clumsy structure. Visual material and use of tables, is not relevant or is inappropriate.

Fail – Unacceptable work (0–2): Unacceptable standard of presentation with poor use of UK English.

Marker's comments:

Marker's initials:
Second marker or moderator comments:

draw out questions and suggestions for more sophisticated discourse. For example, in the passage above from Lidija, the fact that she has shown movement and first-person narration in the third sentence would be noted and pointed out to her as a strong and effective signal that literary travel writing was the discourse she was employing. Two questions to pose to her are to ask if she can make the tense change, 'Just a day later' read more smoothly, and then to look at a keyword in her excellent title, 'Ashes' and ask how can she make this into a twill word; this was explained in depth in Chapter 1. Can she introduce ash again, perhaps replacing frozen with another imagistic word?

Example (2) is an example of personal storytelling as it includes locals, stories and memories related to the writer's grandma. There is also room for posthuman animals, Lucky and Pika. The writer employs languaging, by using the local word 'Svečomat' – an automatic machine that gives candles in exchange for money. As in professional journaling, the day entry begins with a date, a point for which the lecturer can reward Anja during formative or summative feedback. Further, in good travel writing practice, she has verbs of movement very early in the piece, for example, 'take a walk'. She does make a tense shift even within the first sentence, but this should remain since it acts a powerful form of *in medias res* for the day:

> 2    Saturday, 19th March 2022
>
> Lovro, my boyfriend, is still at my place and it is Saturday, so we decided to take a longer walk today. We went past the wood that represents our normal path and went towards the Svetinje church and cemetery. While we walked towards the church we ran into a lot of locals that said hello to us or they were working in the garden. It was warm outside and a lot of people saw the opportunity to get some vitamin D. We continued our path downhill towards the cemetery where my grandma is buried. We bought a candle at machine called 'Svečomat' and lit a candle. We stood by the headstone for a minute and then continued our way uphill towards the church while I told him fun stories from times, when she was still alive. On our way back home few cars passed us and we had to hold Lucky back, because he would just love to run after them. We almost came back to the house when we had to stop again, because Lovro just had to pet and cuddle Pika. (Anja Govedič, My Week, 2022)

In Example (3), the author, Luka, describes the unusually warm weather and beautiful nature. The vivid and detailed description of flowers contributes to the affect of the travel writing text. Two or three very small uses of English send out a signal that it is not a native speaker, 'small talks' in the plural, and 'warmer than I had been counting it would be' rather than 'counting on'. However, it is tempting to leave these unchanged since they still communicate the ideas very clearly and moreover they hint to the readers that they are listening to a local Croatian describing their hometown:

3    19th March 1.00 PM

I was planning on starting my journey around my hometown Samobor (Croatia) on foot. The weather was nice, with the notion that it was still the middle of March. It was a lot warmer than I had been counting it would be, so bringing a bottle of water was not a bad idea at all. /.../ After a few cold months, sun finally came out and birds started their lovely orchestra. Loving sound of the Spring. After a while I came to the main square of King Tomislav and absorbed the heat of warm concrete, which was warming up under the sunlight. Since we all got used to the Winter during past four months, the 18 degrees Celsius was warm. I wanted to avoid any small talks, so the forest surrounding the city seemed much more attractive. Had a little climb, for about 10 minutes. There I sat and enjoyed the fresh air and peacefulness of the forest. I remembered how great it would be to go jogging on the Sava embankment and suddenly started thinking about it. As I was walking back to the city, I noticed a lot of spring heralds in form of different flowers that appear in late Winter or in early Spring. What a joy is to see them! A kaleidoscope of colours. From purple and mix between purple and pink, over yellow to completely white flowers. In this still mostly brown forest, it is lovely to see all those colours covering and decorating the surrounding. (Luka Stanić, Samobor, 2022)

Also, example (4) shows the importance of details, of mentioning particular buildings and streets so that later visitors can find them and re-experience Fran's feelings; notice the affective role of adjectives in travel writing descriptions:

4    Suddenly I began to feel the vibe of Zagreb, and I understood what famous Zagreb poets wrote about in the late nineteenth and early twentieth century. There was no longer so much modern architecture, commercial shops were turned into small family crafts, and modern city lighting was replaced here and there by some old lantern. I decided to get off my bike, tie it to a nearby pole and continue walking. At that moment, I was next to the building of the Croatian National Theatre. Wow. What a building. I walk past it at least three times a week, but I have never stopped and looked at it. The focus on detail in the old architecture is impeccable. Each pillar is decorated with a multitude of small statuettes, and the pane of each window is carved in a different way. It's amazing how much people used to focus more on details. It was enough for me to turn my head to the right and, at a distance of about a hundred meters, to see a modern building that looked like a glass box. No charm, no soul. I was already impressed and overwhelmed. And that was just the first building at the entrance to the older part of town. Then, I headed towards the most famous street in Zagreb, Ilica, and my

next goal was the Zagreb funicular. As I walked along Ilica towards the funicular, I did not allow myself to wander to my daily worries; instead, I focused entirely on the culture and history that was surrounding me. I could feel the smell of the city and hear the murmur of cheerful people from nearby cafes and the passing of the famous Zagreb blue trams. Then, I came to the famous funicular. Definitely not a marvel of technique. Especially nowadays. But that is not the goal. The reason why the people of Zagreb love to ride the funicular so much is not because they do not want to climb the stairs to the upper city. Rather, it is because they feel like their ancestors, who were the proudest people in the world in 1890 when this, then modern, mean of transport adorned their city. As I arrived in the upper town, I felt a heavy nostalgia. Not just because of all the history, but because of the fact that my old high school was standing right in front of me. Unfortunately, due to the earthquake that hit Zagreb two years ago, the school was severely damaged and is currently being rebuilt. Thinking about it made me sad, but at the same time, it made me proud. It made me proud to be a resident of a city that shows the greatest togetherness in the most difficult moments. (Fran Pokasić, Zagreb at Night, 2022)

Example (5) emphasises the local people of a small town, languages spoken, the river and attention to detail.

5    Monday, 21 March 2022
     While walking down the street I saw two elderly women walking towards me while talking. When we passed each other, I kindly greeted them and then they both looked up and greeted back while smiling. I think that a small gesture like that can make another person's days better and it also makes you feel good being a local and giving out a good impression not only of yourself but also about the community as a whole. Observing the city like I was a tourist made me discover some things I didn't see before like some alleyways I never walked down. For me, the interesting thing about my town is the fact that it is located directly on the border with Austria and this makes the people you meet even more diverse. When I came to the café that is located by the Mura River, I could hear people speak German more than I heard Slovene and the fact that this is so is because a lot of Austrian tourists that stay in the nearby Thermal SPA centre come across the border to have a delicious cup of coffee. I think hearing other languages being spoken in your own town makes it so more exciting and it makes you want to go and ask them what they think of the town. (Erik Miklin, 2022)

In example (6), the author describes the most characteristic attraction of a destination and a trip to a vineyard – again with great attention to details from interior design and the light to the description of how wine is produced – all this significantly contributes to persuasiveness:

6    Zreče and its surroundings also have a lot of vineyards and great wines. Me and my boyfriend visited the Andrejc vinery, which is a popular hangout spot for local people as well. They have a lovely tasting room, where the owner takes you through the tastes and sorts of white wines, they produce. Andrejc vinery has a long tradition, that has developed through generations of the Andrejc family. They are very passionate about their work. When I walked into the tasting room, I immediately got a familiar feeling of home, because of the cosy interior design, with wooden elements. The lighting was dim, so it added a special feeling of elegance and professionalism, that you get from the kind people working there as well. The wine they produce varies from sweet to dry, but the most boutique and special is their sparkling wine, which is made in very small quantities. It is made with the classical method, which means they turn the bottles by hand. The sparkling wine is sweet, made from the Moscato grapes. The bubbles just melted in my mouth, and I was left with a sweet aromatic aftertaste, that I just wanted to stay there forever. And the wine cellar. I could just stay in there for hours and relax. Narrow corridors, with Ambiental lighting throughout, as I slowly walked by hundreds and hundreds of bottles. It was no less than magical. (Ina Olup, Explore, Feel and Enjoy – Done, Done and Done – Zreče, Slovenia, 2022)

Example (7) presents an inclusion of ethnobotany ('teloh' – hellebore) in a travel writing text; this was one of the taught suggestions from a choice of what to include. Through dialogue and questions, the author needs to be made aware of the tenses chosen and what effect moving to past tenses will have for her readers:

7    At first, the path is quite nice and easy, wide so two people can easily meet. If we look around, we can see many small spruce trees, and beneath them in the springtime there are growing pink-white, star-shaped, cute, little flowers called 'teloh' or hellebore in English. But please don't pick it up, because it is poisonous. Just look at it and admire its beauty. (Nika Kramžar, A Hike to Lisca, 2022)

Example (8) shows the significance of geographical and historical data and inclusion of senses in travel writing texts:

8    On a sunny spring Sunday, I decided to take a long stroll to a nearby town, a small town if you can even call it that. It is no bigger than any other village, yet it holds great value as a historical place. A few kilometres from my home, the smallest town in Slovenia, Kostanjevica na Krki, is hidden among the mighty Gorjanci hills and vast Krakovski forests on the banks of the emerald beauty of Dolenjska

region, the river Krka. It is a hidden paradise, offering something new and unexplored upon every visit. The walk starts on an old wooden bridge spanning over a calm and slow river. The creaking wood echoed behind me as I crossed the bridge and at my side, my four-legged friend Ron faithfully followed my every step. I continued my walk through the old town centre, which is located on a river island. Yes that's right, a part of the town is an island, situated on the Krka river. Because of its remarkable location on the river, the city gets flooded often, and sometimes the only possible way to get around is by boat. And that is the reason it is also called 'Dolenjska Venice'. The houses on the main street looked very romantic in the early morning, when the first rays of sunshine shone on their lively coloured facades. I stopped by the church, surrounded by a small stone wall. The church bell tolled the hour when I noticed the joyful play of birds on the bare branches of the mighty linden tree that stands in front of the church gate. At first glance, the church seemed small, but a closer look at its bell tower made it seem imposing and magnificent. The patron of this church is St. Jacob. I read that this church is the oldest building on the island and one of the oldest Gothic churches in Slovenia. I noticed how the image of the old town was a mix of traditional and modern architecture, with the facades of the houses telling their own unique story. The town centre is divided into two parallel streets, which nicely encircle the houses between them. I thought to myself, with only two main streets, no one can get lost here even if you don't know where you are. The streets are still known as 'Tamal' and 'Tavelik plac'. The left one, Oražnova street, that leads from north to south, eventually opens the funnel and forms a market ('plac'). In front of the south bridge, the streets merge into Kambič square. The streets are named after the important people and events that marked the city through time. I stopped at a tourist information board in front of Lamut's art Salon and read a few paragraphs about town's history. It filled me with joy to know how this small town influenced the local history Kostanjevica is considered the oldest city in Dolenjska, and its history goes back to thirteenth century. In the early thirteenth century, the Carinthian duke Bernhard von Spanheim established the Cistercian Abbey on the southern frontier of the March of Carniola, which he claimed against the resistance of the Patriarchs of Aquileia and the Dukes of Merania. (Anja Kraševec, Kostanjevica na Krki, 2002)

Waters and plants have an essential role as situated moments in travel writing. Krka is a beautiful river of dark green colour that flows through the almost fairy tale town of Kostanjevica na Krki. Reference to Venice is important in the example above. Also, the name Kostanjevica is interesting as 'kostanj' means 'chestnut tree' in the Slovenian language. Anja's naming of the street will help visitors to re-live the walk. From the naming of the streets and the mention of the church much of this would be classed as travelogue; it would be useful for the lecturer to see the original field notes here to see if any personal feelings could be elicited and used to add more to the story.

Local gastronomy is a topic that travel writers like to mention in their texts. If example (6) mentions wine, example (9) displays the mention of a beer:

9    In the evening I returned to the centre, I met with my friend at the Croatian National Theatre, personally one of the most beautiful buildings in Zagreb. We walked along Ilica – the largest shopping street in Zagreb and commented on how everything has come to life with the arrival of warmer weather. We came to the main square and there were so many people that we barely made it to Tkalčićeva, one of Zagreb's most famous streets for nightlife. After a few 'Grička Witches'- famous beer, we returned home and continued to comment on how great the weather is. (Ivan Jokić, Zagreb, 2022)

Describing an itinerary to a destination makes valuable reading, especially to visitors in search of information on reaching a destination. Example (10) displays a text with some basic instructions on how to get to a destination and what to expect there.

10    I travelled to Sarajevo with my mother for only two days. Her great wish was to visit Sarajevo, so I took her with me. We travelled from Zagreb to Sarajevo by bus and the trip lasted eight hours. We travelled at night and arrived at 6 AM at the bus station. The hotel was quite far away so we took a taxi to get there. As we were tired from the trip we decided to go to bed first before going to breakfast. After a short nap, we went down to the reception desk to meet the woman I was communicating with regarding the hotel tour. While my mom sat in the hotel lobby with tea, I went on a tour of the hotel that didn't last long. After that, the hotel manager treated us to breakfast in a restaurant located near the hotel. The hotel was located in Baščaršija, the most famous street in the city, so we went sightseeing alone because I knew the city well. We had the whole day to ourselves, so I tried to show my mom as much as possible. We first went to the very heart of the street on the so-called 'Pigeon Square' where the famous sebilj is located. Sebilj is a small kiosk in the Islamic architectural tradition where water is freely dispensed to members of the public. There we took pictures and headed towards the symbol of Sarajevo, which is the City Hall – a monument to the multiculturalism of Bosnia. Opposite the City Hall is Sarajevo's iconic cable car, a trip up the mountainside. Shiny new cable car station is in the foothills of Mount Trebević, one of the peaks which played host to events in the 1984 Winter Olympics. For a return fee of 20 Bosnian marks (approximately £10), this must-do

cable car lifts you more than 1,100 m in seven minutes, providing breathtaking views every second of the way. It's a view which defies comparison with most other European cities. Mosques and minarets decorate the skyline along with the Romanesque towers of Catholic churches and the onion-shaped domes of Orthodox ones. And that is another thing which makes this city so fascinating: it's a place where east and west meet. (Marica Ilić, Sarajevo, the City Where East Meets West, 2022)

Students' feedback on the writing task involving travel writing was positive. They commented on the task as follows:

*I always go for a walk with my dog, but this time I had opportunity to write about it. This meant, that I had to pay more attention to the events around me. For the purpose of different stories, I also took some new paths, which broadened my horisons. I love using adjectives in writing and travel writing gives me the chance to do so, because for example in our seminar work or diploma we have to use more professional language. (Anja Govedič)*

*I really enjoyed the writing, rediscovering my hometown and analysing the environment around it. Travel writing is emerging phenomenon in our area (the Balkans and ex YU) and people are starting to recognize the importance of this kind of writing and expressioning. In free time, whenever I remember to read or feel the urge to read, I am enjoying travelogues. Travel writing reminded me of travel experiences of famous world avanturists, travelers, bloggers and made me feel of contributing to someoneones experience and ones imagination. I would like to work on some type of travel writing in my career. Since I am working in tourist agency for unconventional voyages, I am expecting to work as their tour guide and as someone who will enlight the importance of expression during and especially after travels. I would like to combine storytelling of travels and present it on agency`s website. Currently I am editing voyages on website, therefore presenting travel writings of my own and of our travelers, would be beneficial for promotion and for feedback of our field work. With this task I have realised there are different aspects of travel, that we may and can introduce in the firm and business world; not only to view travel writing as academic work, as people might have been seeing it before, but as a way of expressioning to others by inviting them on voyages and sharing experience (co-creation of experience on some level). This could be possible for readers of our e-magazine. This shows multidisciplinarity of tourism and how broad travel writing affect could be. This is not a way of showing travel writing as a mean of promotion, but as a way of tourists expression to others by sharing desire for travel. (Luka Stanić)*

*I have enjoyed the travel writing task very much. Mostly because it was a way of expressing our feelings, since most of the text we usually write at the Faculty are objective. It was a good exercise for testing and expanding our vocabulary and creative writing. Since I am a new temporary resident of Zreče, the task allowed me to experience the town in a different way and it helped me to discover my new home. (Ina Olup)*

*The thing I liked about the travel writing task was that I was able to ex-plore and see my city through the eyes of someone who is just here for a visit. I also think that every person wants to do different and see different things and through travel writing we can share our own point of view of a travel and this makes us realize how different and unique each and every one of us sees the world. Through travel writing we can share the beauty of differ-ent places and countries with people who are unable to go there themselves. (Erik Miklin)*

*I am glad to say that I enjoyed the task and I am very glad that you liked my essay. I feel a great love for my city and I am very emotionally connected to it, so it was a pleasure to put my feelings down on paper. I like to write and unfor-tunately I don't often have the opportunity to write texts that are not written in a formal or scientific style. So in my opinion the task was a complete hit. (Fran Pokasić)*

*I really enjoyed the task travel writing. It brought out my feelings about the hike, I was more observant. It is different way of just hiking, it is hearing, look-ing around, admiring it. My opinion is that we should do the task more often. (Nika Kramžar)*

*This task was very interesting to me, especially the result of the same be-cause I essentially saw my regular day from a different perspective. When a man reads it looks much more "exciting" than it seems live. I think that this task broadens our perception of experience. (Ivan Jokić)*

*Regarding the travel writing exercise, I would like to say that I really en-joyed it. It was a whole new experience to write about my travels and it gave me another perspective on how to be more observant of my surroundings, people, and feelings. I have found that in any travel, or an everyday walk, the jour-ney by itself is more important than the destination and the way I collect my thoughts while writing is quite different for me because it gives a deeper mean-ing to the experience itself. I will carry on with travel write for myself during my next travel. (Anja Kraševec)*

*I liked the task because I was in Sarajevo a few days before I started writ-ing, so my memory was very fresh, so there were no problems with writing. My favorite part of the job at the agency is writing travel plans so I enjoyed this as well. (Marica Ilić)*

*Writing, travel writing was a minor problem for me as I would rather inspire people with a spoken word than written. (Lidija Kočevar)*

### The author-editor on a literary travel piece from Newquay

This case-study section looks at an extract of dialogue between peers, one as the author-editor as a final travel story is put through copy-tasting and final questions for the writer. The reception by six readers of the final text is reproduced to provide a comparison between an editor's meta-language and the readers' language.

*Author's editor on a near-final version of 'Fieldnotes from Fistral'*

> They wake, they work, they wait,
> Then they fall,
> Like the gulls call to the shore:
> Ro an mor, ro an mor.
> from 'Seagift'
> February 2020 had been a hard month, not through cold, no, but from the warnings sent in by the Atlantic. We woke to wet slate on the terraces each morning. On Tuesday 10th March 2020, I headed west to Newquay on the Great Western Railway with my first stop to change at Par. A fine, misting drizzle blew across the fields from the sea outside Plymouth; the promised sunshine of the vernal equinox, now only ten days away, had not yet materialised. After the change of trains at 10:10 AM, onto the Newquay branch line, I crossed Cornwall from south to north. From Par onwards to Quintrell Downs, yellow gorse bushes were already in bloom, usually called furze in the Westcountry. The railway had followed this course into Newquay since 1876. I was not due to meet staff at the museum in Trenance Heritage Cottages until the following morning, so I had the whole afternoon to lunch and to explore Newquay town.

*Dialogue on fistral writing with author's editor*

JPT: It is interesting how you start your narration about February being a hard month, and then jump to 10th March, when you started your journey to Newquay.

CM: Yes, that month change is abrupt and has no causality. Thank you for drawing out that. I think it needs a link. I want to keep the Atlantic because of the surfing later. I need to think longer and make that date transition have a reason.

CM: What do you think of keeping the Atlantic Ocean as an active agent? i.e. 'by the Atlantic' or 'from the Atlantic'?

JPT: Yes, certainly, since you use the Atlantic again when noticing the surfers. Maybe a few words more about Natalie and reasons for visiting? Or you, perhaps, wanted to employ the element of mystery?

CM: No mystery intended, so, yes, it would be better to introduce Natalie here in paragraph 1 rather than just 'staff'. It would introduce the quest element, too, of a named character I was planning to talk to.

*Reception – readers' comments on Fistral writing*

1  I really enjoyed reading it! You've made me want to visit again – alas, I will have to wait until the lockdown is over.
2  [...] this makes me feel so different about it. The luxury of non-essential travel, by train, having fish and chips, coffee, staying in a hotel, breakfast there, and visiting [...] the museum.
3  I could picture all of your journey and [...]

4  I found it much better for helping the reader imagine themselves in a place. Especially with the little details you came across, such as the review in the newspaper, I suppose I would say it made it a more immersive read.

5  I wouldn't need a map to find my way round. It's good how you combine literature, history and food.

6  A wonderful piece, such a read, beautifully put together [...]

To be useful to the writer, readers' comments need to be interpreted as hermeneutic texts. Expressions of enjoyment are easy to understand and encourage the writer to continue. However, more in-depth attention is needed to see that the mention of stopping for food has a powerful affect on the reader.

## Giving feedback on writing

In writing, feedback is essential and challenging at the same time. A teacher is a guide and a feedback provider (Carter & Kumar, 2017). But many students in higher education, who are crucial stakeholders in this process, feel that the feedback process is not as useful as they expect, and they report that they struggle with the information they get from their teachers (Hopfenbeck, 2020). Yet, both teachers and students are crucial in this process that is perceived as a significant link between teaching and learning (Plank et al., 2014).

What is feedback? It seems that everyone is familiar with the term, but concepts behind it may vary. Ramaprasad (1983, 4) defines feedback as 'information about the gap between the actual level and the reference level'. According to Stracke and Kumar (2020, 268), feedback is 'a key element of learning and development for both the supervisor and the candidate'. But perhaps the most precise explanation of feedback is Carless' (2015) who argues that this is a process in which students reasonably employ reviewer's information about their performance and use it to improve the quality of their work or strategies of learning. The latter could be applied to teaching and learning of writing skills as well. It is seen, as Evans (2013) argues, that student represents the central point of the feedback process and that this process is influenced by a student's needs. Or better phrased, the process of feedback should be designed by students' needs and should consider various contexts since it takes place in various contexts and conditions (Henderson et al., 2019). There are different views on the topic whether the feedback should be oral or written, but many authors agree that the design of feedback is significant (Henderson et al., 2019). Bitchener et al. (2010) believe that written feedback is not necessary if supervisors and students meet often enough and discuss the expectations of the content. Many things should be considered, among them circumstances in which instructions are given, characteristics of a task and of course, student individualities (Shute, 2008). Another very important aspect is the quality of feedback. Experience and research show that ideas, concepts and definitions of appropriate feedback may vary greatly (Haughney et al., 2020).

Stracke and Kumar (2020) enumerate the following aspects of feedback: issues with content (and medium), issues with linguistic accuracy, issues with the direction of feedback, issues with the (critical) language used in feedback, issues with emotions in feedback, issues with the supervisory feedback,

Kara (2004) offers the following 12 tips in her online blog:

1 Be honest in all the feedback you give.
2 Read the piece you're giving feedback on carefully, thoroughly, at least twice.
3 While you read, make notes of thoughts that occur to you. As a minimum, these should include: aspects of the work you think are good; where you think there is room for improvement; anything you don't understand; references the author might find helpful.
4 Be sure to praise the good points in the author's work. This helps build trust and also lets the author know what they can relax about.
5 Be open about anything you don't understand. Doing this worries some people because they think they may look stupid, particularly if they're giving feedback to a peer or colleague rather than writing an anonymous review. But it's really helpful feedback for writers because it may be that they haven't written clearly enough.
6 Give a straightforward assessment of areas where you think there is room for improvement.
7 Tell the author *how* you think they can improve their work. This is crucial. If you're only saying *where* improvement is needed, you're only doing half the job.
8 Where relevant, suggest references the author has missed.
9 If you think extra references would be helpful but nothing specific springs to mind, have a quick look on a website such as Google Scholar or the Directory of Open Access Journals and see if you can find something to point the author towards.
10 Don't worry if you can only offer a certain amount of help because of the limits to your own knowledge. It's fine to say, for example, that a quick online search suggests there is more relevant literature in the area of X; you're not certain because X lies outside your own areas of interest but you think it would be worth the author taking a look.
11 Acknowledge the author's emotions. For example, after giving quite critical feedback, you might say something like,

> I realise that implementing my suggestions will involve a fair amount of extra work and this may seem discouraging. I hope you won't be put off because I do think you have a solid basis here and you are evidently capable of producing an excellent piece of writing.
> *(Though remember #1 above and don't say this if it's not true.)*

12  Be polite throughout, even if your review is anonymous. Anonymity is not an excuse for rudeness (Kara, https://helenkara.com/2018/12/04/how-to-give-feedback-on-academic-writing-twelve-top-tips/).

### The meta-language of feedback in writing

The author-editor, the teacher and even writers themselves, when revising their own work, need a vocabulary for pinpointing what is good and what is not in the texts they review. Myhill, Newman and Watson (2020) examine the value of a meta-language for teachers and learners to use when discussing and trying to improve written work. They take the approach that 'teaching develops students' metalinguistic understanding of how written texts are crafted and shaped' (Myhill et al., 2020, 2), and thus grammar and sensibility to language structures can add value to the texts created through dialogue. Grammar in the hands of the writing teacher is 'a fundamentally functionally-oriented perspective on grammar' (Myhill et al., 2020, 3) and hence is not a descriptive nor a prescriptive grammar. The author-editor, therefore, uses grammatical terms, and literary terms, to help the learner focus on a structure. For example, adjectives often precede the noun they qualify in British English, but to sound more archaic or literary, the writer can make the decision to place them after the noun. This draws on the echo of Norman-French being used in courtly settings in England from 1066 until around 1500, in which adjectives follow the noun. The following example is offered from the travel book of Vita Sackville-West as she arrives into Isfahan:

> But all around, in the twilight that encircled that focus of glow and frenzy, the Meidan lay like a lake of peace, long, narrow, level; and within the entrance to the bazaars a single lantern burned, showing the way into that obscure and unfathomable warren.
>
> *(Sackville-West, 2007, 111)*

Her travel writing here, above, increases in emotional intensity to convey her feelings and the beauty of the view across Naqsh-e Jahan Square in the city, the maidan, in modern spelling. To do this, she uses a simile, comparing the square to a lake, and then positions the qualifying adjectives after the noun so that the image of a lake is not lost, and so that the alliteration on the letter 'l' maintains its proximity. Very soon after, in the same scene, she uses adjectives after the noun, 'pleasure' to create a different effect:

> I had no time to go to Shiraz and Persepolis that April, but it was a pleasure deferred, not a pleasure foregone.
>
> *(Sackville-West, 2007, 111)*

Here, arguably, the literary travel writer is activating in the reader's tacit knowledge the title of Milton's (1671) poem *Paradise Regained* since Persian garden design is referred to as the paradise garden. Myhill *et al.* continue and emphasise the problem of a tacit knowledge of the nuances of literary English that has never been articulated and so lacks a vocabulary:

> In the writing classroom, it is difficult to share tacit knowledge about effective writing, because of the inability to verbalise it. Explicit grammatical knowledge, however, is accessible and usable grammatical knowledge: it is 'learning' knowledge, as it can be used to develop greater understanding of how to write, how to solve writing problems, and to share thinking about grammatical choices in writing.
>
> *(Myhill, Newman & Watson, 2020, 6)*

Place-branding, in its analysis phase, shares this same problem when interviewing locals about what they enjoy about their own city; the locals' knowledge is tacit and has never been called on before by a stranger. In writing, the everyday response when asked if a piece of travel writing is good, is to say 'it works well' or 'it's great', and the case study on Newquay, above, reproduces six real feedback comments from the public which demonstrate this linguistic register. Writers who are trying to gain some feedback that they can implement are left at a loss. However, if the author-editor uses grammatical terms, and literary terms to pinpoint where changes can be made then the focus is clear; the changes can be made and a review of the success or otherwise of these changes can be evaluated.

# 8

# CONCLUSIONS, RESEARCH FUTURES AND MANAGEMENT IMPLICATIONS

## Space informatics and the ubiquity of online mapping

Between June and November 2007, Apple made its combined iPhone computer and mobile telephone available in the United States, the United Kingdom, France and Germany. For the current generation in the west, this means that the ubiquity of internet access is as commonplace as being able to carry a book to read. Google Maps, the online mapping web software, is slightly older than the iPhone by 2 years. Bringing the mobile hardware and this software together has meant that users are comfortable by the 2020s with using these as a portable A to Z of any city in which they arrive for leisure. Almost all hotels and many transport companies are improving the way that they provide wifi access to tourists both during their stays and onboard trains, ferries and in transport waiting areas while they travel. The appeal of these wifi connections is both their bandwidth and that they do not incur the roaming charges of cell-phone data transfer. The final piece of this jigsaw is the design of e-commerce interfaces for online travel agents, hotel groups and larger carriers which give tourists the confidence to buy on the move. This was briefly termed m-commerce. The result of this convergence of these technologies creates a new mentality for the tourist-buyer. At least, for short-haul travel, they feel confident in not having all their accommodation booked in advance, which means that this new tourist can follow a more flexible itinerary during their time away from home. They are more exploratory and can follow initiatives or desires created even during the holiday.

An area of an unknown town, or the neighbouring town to the tourists' current base, is less closed and less threatening. With mobile online mapping, these hitherto hidden spaces are available for tourists to make more personal choices based on their emotion at the time. A more fluid

DOI: 10.4324/9781003178781-9

communication text is needed to reach these decision-makers to help them find and explore places that will suit their experiential needs. Travel writers with their partner DMOs will need to unflatten the traditional map and with geoinformatics tell holidaymakers that a place of great value to them is only just a few streets east of where they normally walk down to the beach, for example.

The What3words initiative to convert Cartesian x and y coordinates, the eastings and northings of the paper map into understandable and memorable speech was a success but still did not handle the z-axis. The travel writer can tell stories of what happens in the basement or how to find the RER station Magenta deep underground in the labyrinth of the Gare du Nord in Paris, somewhere below a house on a street at 48.88009497942813°N and 2.3582571744918823°E. Or on which floor to find the South Memory restaurant in the Grand Gateway commercial centre in Shanghai. Along with a more three-dimensional description of place, written narrative can re-tell an experience of finding and being on that level of that space in a way that readers can process and decide if they want to re-live the moment there.

## Journaling technologies

Web 2.0 technologies, where users can write to servers in the cloud and can share those data across their own devices and with selected other users, has given leverage to journaling for writers working on a project basis. The zettelkasten design for use in Microsoft's OneNote described in this book provides a valuable working model for professional writers. Journaling relies on dialogue, initially with the source documents for a project, then with field experiences and with the stakeholders who will be users and readers of the final pieces. Having an online, always ready, central repository for the notes that the writer makes, means that ideas need never be lost. In addition, with a structured capture format, the catalyst for any new enquiry can also be stored to consult for the detail needed during the write-up. With Microsoft's investment in the product, the journal-keeper is assured of a robust user interface that makes working a pleasure. This ease of use, or enjoyable user interface, is vital since the journal is not just a place of rapid capture but must become a space for constant re-working, editing and addition of new material. In addition, during filing, the user must feel at ease in creating cross-reference links to other pages across their journal so that it becomes their reference book for the current project. It demands discipline, and poor usability features will interfere with this daily logging, refining and filing. If maintained, the search function provided will quickly take the writer back to a point they need from earlier in their journaling.

Web 2.0 means that the project's lead writer can give access to the shared zettelkasten to co-writers, mentors and, of course, stakeholders. All of these collaborative users can stimulate the necessary dialogue that generates new knowledge in a processual manner. A co-writer might ask the lead writer to explain what they mean, or a stakeholder can be asked a question and leave their answer in the specific OneNote page that discusses an idea or a place in the urban space. Leaving the working replies within the OneNote notebook means that they can be returned to at a later date without the concern that the whole journal will become too large and impossible to navigate. This is often an issue with paper-based card indexes. It is usually true that some of your key stakeholders are also your buyers, commissioners and readers. Having this almost live relationship with this sub-group during the writing process enables the skilled writer to fill gaps in the finished piece and thus produce a final story that better meets the readership's requirements and values. In purely contractual and project management terms, too, the commissioner can see progress. In fact, the template design is built upon prototyping interim deliverables along the critical path of the writing project.

The development of handwriting recognition software, though, has rescued the notepad and index card. Many times, it is necessary to keep notes by hand, especially on transport systems and in the field but up until recently, this meant painstaking copy-typing to merge offline data. Now, with a little care and having ruled paper, as the research for this book discovered, handwritten notes can be scanned, transcribed by software and then incorporated into the online journal. Interrogating destinations by writing means that the more the researcher can convert ideas into writing on-site, rather than taking photos as memos, the more immediate the writing inquiry will be. A temptation exists to photograph the destination and then postpone the inquiry by writing until later, off-site, but, of course, this serves to interrogate the photographs rather than the place.

Dictation software is available at the time of writing but initial trials created too much interference with the field research process. However, it may be that the technology will quickly become more suited to the research of place and space, and the researchers will themselves develop more skill with handling the recordings they make.

## User-generated content in the travel industry

Advances in technology and social media have enabled platforms to be inclusive. Contemporary platforms, thus, allow widespread dissemination, stimulate creativity and active citizenship and play a key role in the development of communities and economies since they have the ability to bring many stakeholders together (Gillespie, 2010), to empower citizens to become active users, to express themselves and become cultural producers in their own right (Duffy et al., 2019). Especially in tourism and in the travel industry, user-generated content contributes significantly to the quality of services through suggestions, judgements and post reviews (Kitsios et al., 2022). Users not only create and disseminate their experiences, but more recently, they also have the possibility to interact with other users and share information created by other people (Li et al., 2019). Experts (Aydin,

2020; Kar et al., 2021) attribute this crucial dialogic role of user-generated content to the fact that tourists, travellers and users of services trust the information of users and want to know first-hand a personal, emotional assessment of the quality of a particular service. Recommendations by people who have already visited a certain destination or experienced services they review (experiences may present a combination of ethos and logos in Aristotle's taxonomy of means of persuasion) represent valuable sources of information to people planning their visits and travelling (Assaker, 2020; Kitsios & Kamariotou, 2021). These user-generated data and information, in fact, influence consumer behaviour in the hospitality sector (Kim & Kim, 2020; Xu et al., 2021) and contribute to improved decision-making in the tourism sector (Kitsios et al., 2021; Kitsios et al., 2022). Consequently, the role of user-generated content in destination image and destination development has become essential. And when it comes to long-term relationships with customers and tourists, according to Kitsios and Kamariotou (2021), perceived value on social media platforms (Facebook, Twitter) is the most important factor influencing potential consumers' trust.

Large amounts of information require selection (Greco, 2018) on the one hand and good persuasion techniques or strategies on the other. In persuasion in the tourism sector and in general, storytelling techniques are vital. Among the three means of persuasion (ethos, pathos and logos, according to Aristotle), pathos as a means of persuasion is particularly important for engaging audiences. Writers, reviewers and other content creators need to show their enthusiasm and passion about places, destinations and services because this is what creates emotional connection with audiences and consequently contributes to more successful promotion (Cheng et al., 2020). Innovations in artificial intelligence and in virtual reality provide important opportunities for customers and users to create various contents that influence markets (Van Laer et al., 2019) and improve destination marketing campaigns (Yousaf, 2022) but these still need storyboarding with strong narrative elements including character and plot. Billions of euros, pounds and dollars are invested into digital stories which have consequently developed into the huge and significant industry of content marketing (Van Laer et al., 2019).

The use of social media creates numerous opportunities and enables immediate dissemination of impressions, which further enhances new actions and opportunities. The potential of mobile apps, online social platforms, blogs, online audio and video streams has been recognised by relevant business, administrative and educational institutions. New media is effectively used in promotion, marketing, advertising and branding, and it is due to social media that consumers have taken on both the role of users and producers of information (Dettori et al., 2016). By using advanced technology word-of-mouth, WOM and eWOM, messages can be transmitted anywhere and anytime (Blaće et al., 2015, 113). Since consumers have become influencers, the perception of a place or destination is crucial. The use of social networks is essential for enabling people to share their experiences (Xiang

& Gretzel, 2010). All the media, especially the digital ones, have their audience through which they influence the assessment of the media content. A positive word of mouth creates a favourable image, and the negative word of mouth may have negative effects on the image, perception and potential customers' intentions to visit an attraction or a destination (Zhang et al., 2014). User-generated content in the travel industry is significant because travellers very often choose services, destinations and accommodation by themselves based on the experiences shared by the social media users.

Due to accelerated contemporary life pace, a faster and faster exchange of information is demanded, and that is why mobile applications and social media are continually updated and newer models are bought by existing users. With more secure and powerful mobile computing in phones and tablets, increasing numbers of tourists choose the accommodation only when they are arriving at the destination and they use mobile applications to book the accommodation on the move (Potočnik Topler, Zekanović-Korona, 2018). Research shows that digital contents feature a positive influence on tourism and that at some destinations where research had been carried out, these contents resulted in a significant increase in number of visits and the number of bed-places. One of the examples is Croatian Zadar, which was voted European Best Destination in 2016 (Potočnik Topler, Zekanović-Korona, 2018). Based on this, it can be concluded that destinations, attractions and travel services should be carefully described on digital and social media to ensure that visitors are not misled by advertising claims that have not been properly researched. Digital media campaigns, in fact, are fundamental, but this means investing a significant amount of money in order to gain any return on investment. In addition, marketers can follow users' reports on social media channels in order to see where to make improvements in the tourism offer and to achieve greater tourist satisfaction. Those companies and DMOs that have invested in this type of observatory practice create the environment for their smart destination specialisation and to upgraded sustainability practices in their urban spaces.

## Environmental ethics and posthumanism

Local communities and small villages away from the busy city centre or quayside tourism centres are very important for hosting small-scale attractions and small events despite the fact that in the past tourism managers often forgot about the stakeholders in remote areas (Koščak & O'Rourke, 2019). The inclusion of the locals in the development of a destination is essential. If the locals and experts are not actively involved in the environmentally aware strategic planning, many negative impacts may emerge – from heritage sites turning into theme parks that are overrun with tourists (Koščak & O'Rourke, 2019), to devaluation of heritage and loss of authenticity. Discussing and planning tourism with the locals is demanding, but involving schools, local entrepreneurs, farmers, associations, district representatives, along with policy-makers is a precondition for success in sensitive development. Constant communication with all the local stakeholders with a focus on the locals' interests in the process of preparing a destination development strategy and also later when the development strategy is put to practice is the foundation of developing successful sustainable destinations, with a neutral and low-carbon footprint. Place-writing skills, as shown earlier, are essential for this communication process. Learning to write about urban space and tourism space will assist the clear communication in these projects. It is significant that planning and designing sustainable tourism is an ongoing process and this book has shown how new writing projects can also be designed to be a process to accompany this development. The debates about the positive and negative effects of tourism, sustainability issues, responsible development and active participation of the locals so that the consensus about tourism development is reached are more readily achieved through a processual approach. The challenge at many destinations remains how to create attractive and interesting attractions that are accessible and distributed around their underdeveloped areas to avoid over-tourism in some areas and encourage economic viability in the others. One of the possible tools for attracting visitors and tourists is travel writing, which can be offered for engaging tourists, who are provided with literary texts that have discoverable places, and in offering writing activities to visitors as part of their leisure experience. These places can be lakes, rivers, trees, plants, museums, statues and important built heritage. Well picked stories and their re-telling will engage with visitors through narrative knowing rather than simply through information.

Travel writing tours are one of the most popular ways to connect literature, architecture and culture. To be successful in this, it is crucial to educate and empower experts at destinations to cooperate. This could be done through education, workshops and project activities in local communities. For designing new sustainable tourism products, such as travel writing tours, it is essential that experts from all involved fields work with the locals, who may be encouraged to engage into life-long learning activities that contribute to locals' life-satisfaction, well-being and living a low-carbon life.

Also in this book, travel writing is perceived as a tool for achieving various goals. Insightful texts offer many perspectives, encourage creativity, enhance brand and destination images and contribute to awareness about significant social and environmental topics, including ecology, climate change, migration and identity issues.

## Futures in higher-level education of writers

Graham (2019, 288) argues that

> effective writing instruction involves (a) Writing frequently for real and different purposes; (b) Supporting students as they write; (c) Teaching the needed writing skills, knowledge, and processes; (d) Creating a supportive and motivating writing environment; and (e) Connecting writing, reading, and learning.

The systematic production of personal travel writing serves as a pedagogically sound teaching method and as a tool for developing communication skills with an emphasis on writing and sensitivity. Besides contributing to further development of writing skills, travel writing encourages a non-sedentary lifestyle with the fieldwork implied in the processual method described in this book. Particularly on business and tourism degree and postgraduate programmes, travel writing can be a valuable addition to graduates' competencies. Due to the travelling component, it is usually also quite appealing to students and doctoral researchers, and based on the experience at the University of Maribor, travel writing projects are among the most popular assignments. In the process of researching and learning during English lessons – in groups of English for Tourism at levels from B1 to C1, it became apparent that travel writing is not only a tool for branding places, attractions, products and destinations but also a tool for self-exploration and identity development. Of course, teaching English for Tourism has some specifics, as does teaching English for other specific purposes.

It is fundamental that travel writing is introduced to students with a well-planned theoretical introduction, accompanied by discourse theories and storytelling, which enable better understanding of the process and equip the students with the necessary knowledge and theoretical depth. In turn, this enables the self-confidence in learners to pursue the production of their own travel writing texts knowing that they have a sound methodological basis for testing against research aims and improvement. Despite the fact that in travel writing stories creativity is significant, it is, on the other hand, very important that travel writing maintains credibility by observing some of the requirements, such as: Obtaining reliable information, supporting evidence of what is stated in the text and statements by relevant people, that is, interviews with the locals and stakeholders. One of the essential requirements when preparing for field work within the processual practice of travel writing described in this book, is thorough research. The term, thorough research, here, includes time in archives and libraries, and, in addition to that, studying maps, interviewing experts, reading literary texts, book reviews and following up urban changes through dialogue with city councils.

Developing modules in travel writing is innovative, in the sense that it provides innovative approaches and methods for teaching and learning writing and other communication skills within higher education institutions. Another important aspect of including travel writing into teaching and learning at universities is that travel writing enables interdisciplinary learning and researching, since it connects many scientific fields from Geography, Agriculture, Ethnography, Destination Management, Informatics, to Languages and Literature. In addition, travel writing enables research freedom and encourages active life and creativity, which, consequently, leads to self-development. Thus, travel writing as a teaching and learning method can be placed into many modules, from Linguistic Studies to Destination Development and Content Marketing.

## Building knowledge management platforms for cooperation between public, locals, stakeholders and regional governments

By designing platforms, which are often employed as tools for promoting sustainable tourism practices, destination management organisations, museums and other cultural institutions are seeking methods of presenting destinations and products, revealing the intangible cultural heritage of local residents and sharing this heritage before it is lost. Web 2.0 platforms are very practical tools not only for sharing specific knowledge but also for planning cooperation and tourism development projects because so many users already have the technology skills to participate. On them, stories, experiences, travelogues and reviews are published and available for readers worldwide. However, the platforms are not only tools for publishing and reading but also tools for interaction and mobility, a place for engaging various stakeholders from local citizens and tourism product designers to customers in the process of local place-making.

One of the possibilities of creating platform content is through cooperative research in literary travel writing, through which platform content will enable various participants to see all the written materials and their genealogy. Institutions and DMOs can use these platforms strategically to attract visitors and tourists and at the same time help locals engage in creating authentic cultural heritage.

Organising courses in literary travel writing and place-making to equip local tourism professionals with the ethnographic and specialist writing skills would be a significant next step to developing and valuing new regions and towns. Through Web 2.0 platforms, these professionals can encourage and elicit tacit knowledge from colleagues in the heritage industry as well as local interest groups, so that a cooperative writing practice emerges in the way that this book has documented. Again, posthumanist approaches, especially in writing practices, reappear when cooperative writing using Web 2.0 technologies is considered. Posthumanist writing is marked by finding ethical methods of allowing non-human agents, for example, animals and plants, to have a voice in the constructed text. A further consideration is a new type of text, with a new style or new rhetoric that does not attempt to rationally fix knowledge, but rather seeks for affect in the readers and into a wider sphere:

> If a writing formation or series of words seems unclear or nonsensical, they may be intentionally violating our ideas of clarity and sensibility in an effort to write the new. But by using unfamiliar forms of writing, a posthumanist style focuses attention on how writing – words, formations of words, grammatical structures, and so on – could be different. This is a distinct and subtle point. In the horizon of a posthumanist style, what is important is that writing formations can be different and can be repeated.
>
> *(Rice, 2015, 221)*

Anarchiving, in posthumanist writing, is a way of continually opening up to the future possibility of what a collection of journal entries and dialogue between agents in a shared working environment might be able to affect in the world. Left-over footnotes might eventually have no meaning but Derrida's notion of the spur, Rice (2015, 223) says, reminds readers that they must always take into account that some fragments of the text will always remain without an explanation. This removes or withdraws parts of the text from the hermeneutic field; the reader is relieved of the necessity to interpret this part rationally; instead, sections can be left to create affect. The Web 2.0 platform provides a locus of operation for regional tourism clusters to work as a network of actors (Sair & Slilem, 2022), providing them with common objectives for entering into dialogue and sharing intangible cultural resources. Literary texts have long formed a common heritage around which discussion leads to cohesion in a social group that shares the same heritage; the shared platform extends this cohesion around new heritage created as part of the emerging culture in a region. This has the powerful effect of establishing a place-brand that contains contemporary elements of the destination as it is truly perceived by local actors.

The fragments for opening an archive may be gleaned from research theses, shopping lists, emails, place reviews as well as from deliberate journaling. For example, here in a learned journal article on Deleuze and writing as a form of research inquiry, a fragment of travelogue emerges describing the urban space of Hanoi:

> Riding towards the city centre, I pass some of the rash of mega real estate developments appearing around Hanoi: Royal City and the Vincom Mega Mall on Nguyễn Trãi, Mandarin Garden on Hoàng Minh Giám, The Manor and The Garden on Mễ Trì. None, or so it seems, have Vietnamese names. Royal City and its adjoined Vincom Mega Mall stand like a monument to the excessive wealth of Vietnam's first and only bona fide billionaire, Phạm Nhật Vượng, who made his fortune producing instant noodles in Ukraine.
>
> *(Bright, 2018)*

As each new fragment, like this one, is accumulated in the Web 2.0 knowledge system, it causes diffractions to disrupt the fragments already gathered. The writer returning to synthesise after each new addition will make different judgements and choices as a new city erupts from the corpora.

# REFERENCES

Agarwal, S., & Shaw, G. (2018). *Heritage, Screen and Literary Tourism*. Bristol: CVP.

Agueda, B. (2014). Urban restructuring in former industrial cities: urban planning strategies. *Journal Territoire en Mouvement,* Lille: University of Lille. 23, 3–14. doi: 10.4000/tem.2527.

Alegre, J., & Garau, J. (2010). Tourism satisfaction and dissatisfaction. *Annals of Tourism Research*, 37(1), 52–73.

Althusser, L. (1971). Ideology and ideological state apparatuses. In B. Brewster (Trans.). *Lenin and Philosophy, and Other Essays* (pp. 121–173). London: New Left Books.

Alu, G., & Hill, S.P. (2018). The travelling eye: reading the visual in travel narratives. *Studies in Travel Writing*, 22(1), 1–15. doi:10.1080/13645145.2018.1470073

Anderson, L. (2006). Analytic autoethnography. *Journal of Contemporary Ethnography*, 35(4), 373–395.

Andrades, L., & Dimanche, F. (2014). Co-creation of experience value: a tourist behaviour approach. In M. Chen & J. Uysal (Eds.), *Creating Experience Value in Tourism* (pp. 95–112). London: CABI. doi:10.1079/9781780643489.0095

Anholt, S. (2002). Foreword. *Journal of Brand Management* 9(4), 229–239.

Ansell-Pearson, K. (2010). Deleuze and the overcoming of memory. In S. Radstone & B. Schwarz (Eds.), *Memory: Histories, Theories, Debates* (pp. 161–174). New York: Fordham University Press.

Antosa, S. (2008). Formal and thematic influence of travel writing on the novel: from the pilgrims' accounts to Gothic fiction. *Fogli di Anglistica*, 2(3–4), 55–66.

Aristotle (trans. Butcher, S. H.). (2008). *The Poetics of Aristotle.*

Assaker, G. (2020). Age and gender differences in online travel reviews and user-generated-content (UGC) adoption: extending the technology acceptance model (TAM) with credibility theory. *Journal of Hospitality Marketing & Management*, 29(4), 428–449.

Austen, J. (1818). *Persuasion*. London: John Murray.

Austen, J. (1814). *Mansfield Park*. London: Thomas Egerton.

Austin, L. (2007). Aesthetic embarrassment: the reversion to the picturesque in nineteenth-century English tourism. *ELH*, 74(3), 629–653.

Aydin, G. (2020). Social media engagement and organic post effectiveness: a roadmap for increasing the effectiveness of social media use in hospitality industry. *Journal of Hospitality Marketing & Management*, 29(1), 1–21.

Azevedo, A. (2009). *Are You Proud to Live Here? A Residents Oriented Place Marketing Audit (Attachment, Self-esteem and Identity)*. Braga, Portugal: Minho University.

Barthes, R. (1988). *Roland Barthes by Roland Barthes*. London: Macmillan.

Barthes, R. (1983). *Empire of Signs*. New York: Hill & Wang.

Bassano, C., Barile, S., Piciocchi, P., Spohrer, J. C., Iandolo, F., & Fisk, R. (2019). Storytelling about places: tourism marketing in the digital age. *Cities*, 87, 10–20.

Beaufort, A. (2004). Developmental gains of a history major: a case for building a theory of disciplinary writing expertise. *Research in the Teaching of English*, 39(2), 136–185.

Bechmann Petersen, A. (2006). Internet and cross media productions: case studies in two major Danish media organizations. *Australian Journal of Emerging Technologies and Society*, 4(2), 94–107.

Belsey, C. (1980). *Critical Practice*. London: Routledge.

Benjamin, W. (2019). *The Storyteller Essays*. New York: New York Review Books.

Benur, A. & Bramwell, B. (2015). Tourism product development and product diversification in destinations. *Tourism Management*, 50, 213–224.

Bernstein, J. (1995). *Recovering Ethical Life*. London: Routledge.

Bertella, G. (2014). The co-creation of animal-based tourism experience. *Tourism Recreation Research*, 39(1), 115–125.

Bitchener, J., Basturkmen, H., & East, M. (2010). The focus of supervisor written feedback to thesis dissertation students. *International Journal of English Studies*, 10(2), 79–97. doi:10.6018/ijes/2010/2/119201

Blanchot, M. (1998). The Sirens' song. In L. Davis (Trans.), *The Station Hill Blanchot Reader* (pp. 443–450). New York: Station Hill.

Blaće, D., Ćorić, G., & Jurić, B. (2015). Branding the city of Šibenik as a sustainable tourist destination using social networks. *Ekonomski vijesnik*, 28(Special Issue), 109–125.

Bourdieu, P. (1977). *Outline of a Theory of Practice*. Cambridge: Cambridge University Press.

Bourdieu, P., & Passeron, J. (1990). *Reproduction in Education, Society and Culture* (Theory, Culture and Society Series). London: Sage.

Braidotti, R. (2019). *Posthuman Knowledge*. Cambridge: Polity.

Brewer, W. F., & Lichtenstein, E.H. (1982). Stories are to entertain: a structural-affect theory of stories. *Journal of Pragmatics*, 6, 473–486.

Bright, D. (2018). Writing posthumanist subjects. *Qualitative Inquiry*, 24(10), 751–758.

Brown, G. (2005). Mapping spatial attributes in survey research for natural resource management: methods and applications. *Society and Natural Resources*, 18(1), 17–39.

Bruhn, M., Schoenmueller, V., & Daniela, S. B. (2012). Are social media replacing traditional media in terms of brand equity creation? *Management Research Review*, 35(9), 770–790.

Brunt, P. (1997). *Market Research in Travel and Tourism*. Oxford: Butterworth Heinemann.

Buonincontri, P., & Micera, R. (2016). The experience co-creation in smart tourism destinations: a multiple case analysis of European destinations. *Information Technology & Tourism*, 16, 285–315. doi:10.1007/s40558-016-0060-5, 15.6.2022.

Buonincontri, P., Morillo, A., Okumus, F., & van Niekerk, M. (2017). Managing the experience co-creation process in tourism destinations: empirical findings from Naples. *Tourism Management*, 62, 264–277.

Burbules, N. (1993). *Dialogue in Teaching: Theory and Practice*. New York: Teachers College Press.

Busby, G., & Klug, J. (2001). Movie-induced tourism: the challenge of measurement and other issues. *Journal of Vacation Marketing*, 7(4), 316–332.

Campos, A. C., Mendes, J., Oom do Valle, P., & Scott, N. (2018). Co-creation of tourist experiences: a literature review. *Current Issues in Tourism*, 21(4), 369–400. doi:10.1080/13683500.2015.1081158

Carla. (2015). Goodreads Book Reviews. https://www.goodreads.com/review/show/1252948719?book_show_action=true&from_review_page=1, 23. 04. 2022.

Carlson, A. (2002). Environmental aesthetics. In B. Gaut, & D. Lopes McIver (Eds.), *The Routledge Companion to Aesthetics*, 2nd Edition (pp. 541–555). London and New York: Routledge.

Carter, S., & Kumar, V. (2017). 'Ignoring me is part of learning': supervisory feedback on doctoral writing. *Innovations in Education & Teaching International*, 54(1), 68–75. doi:10.1080/14703297.2015.1123104

Chang, T. C. (1997). From "instant Asia" to "multi-faceted jewel": urban imaging strategies and tourism development in Singapore. *Urban Geography*, 18(6), 542–562.

Chaxel, S., Fiorelli C., & Moity-Maïzi, P. (2014). Les récits de vie: outils pour la compréhension et catalyseurs pour l'action. *Interrogations*, 17. L'approche biographique.

Cheng, Y., Wei, W., & Zhang, L. (2020). Seeing destinations through vlogs: implications for leveraging customer engagement behavior to increase travel intention. *International Journal of Contemporary Hospitality Management*, 32(10), 3227–3248.

Charmaz, K. (2014). *Constructing Grounded Theory*. Thousands Oaks, CA: Sage.

Coates, C. (2014). All alive on the Loire: art, architecture and elephants in Nantes. In *Mail Online – Travel Mail*. London: DMG Media Ltd. http://www.dailymail.co.uk/travel/article-2539767/Francecity-breaks-Nantes-French-fantasy.html, 25.6.2022.

Coghlan, A., & Filo, K. (2013). Using constant comparison method and qualitative data to understand participants' experiences at the nexus of tourism, sport and charity events. *Tourism Management*, 35, 122–131.

Culbert, J. (2018). Theory and the limits of travel. *Studies in Travel Writing*, 22(4), 343–352.

Dann, G. (1996). *The Language of Tourism: A Sociolinguistic Perspective*. Wallingford: CAB International.

Dann, G. (1999). Writing out the tourist in space and time. *Annals of Tourism Research*, 26, 159–187.

Darling-Hammond, L., Flook, L., Cook-Harvey, C., Barron, B., & Osher, D. (2020). Implications for educational practice of the science of learning and development. *Applied Developmental Science*, 24(2), 97–140. doi:10.1080/10888691.2018.1537791

Delcroix, C., & Inowlocki, L. (2008). Biographical research as a cognitive and practical approach for social workers: an interview with Catherine Delcroix. *Forum: Qualitative Social Research*, 9(1), Art. 60.

Dettori, A., Giudici, E., & Aledda, L. (2016). Sharing experiences in tourism: what role can social media play? XXVIII Sinergie Annual Conference Referred Electronic Conference Proceeding Management in a Digital World. Decisions,

Production, Communication ISBN 97888907394-6-0, 9–10 June 2016, University of Udine (Italy), doi:10.7433/SRECP.FP.2016.04

Di, F., Yang, Z., Liu, X., & Ma, Z. (2010). Estimation on aesthetic value of tourist landscapes in a natural heritage site: Kanas National Nature Reserve, Xinjiang, China. *Chinese Geographical Science* 20, 59. doi:10.1007/s11769-010-0059-3

Dixon, P., Bortolussi, M., Twilley, L., & Leung, A. (1993). Literary processing and interpretation: toward empirical foundations. *Poetics*, 22, 5–33. doi:10.1016/0304-422X(93)90018

Dore, L. & Crouch, G. (2003). Promoting destinations: An exploratory study of publicity programmes used by national tourism organisations. *Journal of Vacation Marketing*, 9(2), 137–151.

D'Orey, F., Cardoso, A., & Abreu, R. (2019). "Tourist' sense of place", an assessment of the sense of place in tourism studies: the case of Portugal. *Academy of Strategic Management Journal*, 18, 1.

Dowling, W. (2011). *Ricœur on Time and Narrative—An Introduction to Temps et récit*. Notre Dame Indiana: University of Notre Dame.

Duffy, E. B., Poell, T., & Nieborg, D. B. (2019). Platform practices in the cultural industries: creativity, labor, and citizenship. *Social Media + Society*, 5(4), 1–8. https://journals.sagepub.com/doi/full/10.1177/2056305119879672, 26. 5. 2022.

Dunford, M. (2013). *Friuli-Venezia Giulia Rough Guides Snapshot Italy* (includes Trieste, Aquileia, Grado, Gorizia, Udine and Cividale del Friuli). London: Rough Guides APA Ltd.

Dwyer, L., Chen, N., & Lee, J. (2019). The role of place attachment in tourism research. *Journal of Travel & Tourism Marketing*, 36(5), 645–652.

Ekblaw, E. (1937). The attributes of place. *Journal of Geography*, 36(6), 213–220.

Ellis, C., and Bochner, A. (2000). *Autoethnography, personal narrative, reflexivity: Researcher as subject*. Communication Faculty Publications. 91. Tampa: University of South Florida.

Eshuis, J., Klijn, E., & Braun, E. (2014). Place marketing and citizen participation: branding as strategy to address the emotional dimension of policy making. *International Review of Administrative Sciences*, 80(1), 151–171.

Evans, C. (2013). Making sense of assessment feedback in higher education. *Review of Educational Research*, 83(1), 70–120.

Ferguson, S. (2018). *Diaries Real and Fictional in Twentieth-Century French Writing*. Oxford: OUP.

Fialho, O., & Kuzmicova, A. (Reviewing editor). (2019). What is literature for? The role of transformative reading. *Cogent Arts & Humanities*, 6(1). doi:10.1080/23311 983.2019.1692532, 13.1.2021

Figurski, T. J. (1987). Self-awareness and other-awareness: the use of perspective in everyday life. In K. Yardley & T. Honess (Eds.), *Self and Identity: Psychosocial Perspectives*. Chichester: Wiley.

Forgas, J. P. (1981). Affective and emotional influences on episode representation. In J. P. Forgas (Ed.), *Social Cognition: Perspectives on Everyday Understanding* (pp. 165–180). London: Academic Press.

Forsdick, C. (2009). Introduction: contemporary travel writing in French: tradition, innovation, boundaries. *Studies in Travel Writing*, 13(4), 287–291.

Foucault, M. (1983). Self writing. *Dits et écrits*, 4, 415–430.

Foucault, M. (1970). *The Order of Things: An archaeology of the Human Sciences*. London: Routledge.

Foucault, M. (1966). *Les mots et Les Choses*. Paris: Gallimard.

Friedland, R., & Boden, D. (1994). *NowHere: Space, Time and Modernity*. Berkley: University California Press.

Gale, K. (2018). *Madness as Methodology*. London: Routledge.

Gale, N., Heath, G., Cameron, E., Rashid, S., & Redwood, S. (2013). Using the framework method for the analysis of qualitative data in multi-disciplinary health research. *BMC Medical Research Methodologies*, 13(117), 1–8.

Gentile, R., & Brown, L. (2015). A life as a work of art: literary tourists. *Motivations and Experiences at Il Vittoriale degli Italiani*, 6(2), 25–47.

George, B. (2012). Past visits and the intention to revisit a destination: place attachment as the mediator and novelty seeking as the moderator. *Journal of Tourism Studies*, 15(2), 37–50.

Gide, A. (1925). *Les Faux-Monnayaeurs*. Paris: Gallimard.

Gide, A. (1895). *Paludes*. Paris: Librairie de l'art.

Gieryn, T. (2000). A space for space in sociology. *Annual Review of Sociology*, 26, 463–496.

Gillan, A. (2021). Eat Nantes. In *National Geographic Traveller*. London: APL Media. http://www.natgeotraveller.co.uk/destinations/europe/france/eat-nantes/, 25. 6. 2022.

Gillespie, T. (2010). The politics of "platforms." *New Media & Society*, 12, 347–364. doi:10.1177/1461444809342738

Ginting, N. & Wahid, J. (2015). Exploring Identity's Aspect of Continuity of Urban Heritage Tourism. *Procedia - Social and Behavioral Sciences*, 202, 234–241.

Gladwell, N. & Wolff, R. (1989). An assessment of the effectiveness of press kits as a tourism promotion tool. *Journal of Travel Research*, 27, 49–51.

Godsil, R., & Goodale, B. (2013). *Telling Our Own Story: The Role of Narrative in Racial Healing*. Battle Creek, MI: W. K. Kellogg Foundation.

Gradišnik, B. (2001). *Strogo zaupno na Irskem*. Ljubljana: Debora.

Graham, S. (2019). Changing how writing is taught. *Review of Research in Education*, 43(1), 277–303. doi:10.3102/0091732X18821125

Greco, F. (2018). The rule of persuasion in the marketing process. *Psychology and Behavioral Science International Journal*, 9(5). doi:10.19080/PBSIJ.2018.09.555775, https://juniperpublishers.com/pbsij/pdf/PBSIJ.MS.ID.555775.pdf, 13.6.2022.

Grenni, S., Horlings, L., & Soini, K. (2020). Linking spatial planning and place branding strategies through cultural narratives in places. *European Planning Studies*, 28(7), 1355–1374.

Gretzel, U., Fesenmaier, D. R., & O'Leary, J. T. (2006). The transformation of consumer behaviour. In D. Buhalis & C. Costa (Eds.), *Tourism Business Frontiers: Consumers, Products and Industry* (pp. 9–18). Oxford: Butterworth-Heinemann.

Gustafson, P. (2001). Meanings of place: everyday experience and theoretical conceptualizations. *Journal of Environmental Psychology*, 21(1), 5–16.

Haskoning D. (2020). *Update on the Current state of Knowledge on the Environmental Impacts of Offshore Wind Farms*. Exeter: Haskoning DHV.

Haughney, K., Wakeman, S., & Hart, L. (2020). Quality of feedback in higher education: a review of literature. *Education Sciences*, 10(3), 60. doi:10.3390/educsci10030060

Heller, M., Jaworski, A., & Thurlo, C. (2014). Introduction: sociolinguistics and tourism—mobilities, markets, multilingualism. *Journal of Sociolinguistics*, 18(4), 425–458.

Henderson, M., Phillips, M., Ryan, T., Boud, D., Dawson, P., Molloy, E., & Mahoney, P. (2019). Conditions that enable effective feedback. *Higher Education Research & Development*, 38(7), 1401–1416. doi:10.1080/07294360.2019.1657807

Herbert, D. (1995). *Heritage, Tourism and Society*. London: Pinter.

Hidalgo, M., & Hernandez, B. (2001). Place attachment: conceptual and empirical questions. *Journal of Environmental Psychology*, 21, 273–281.

Holland, P., & Huggan, G. (2000). *Tourists with Typewriters: Critical Reflections on Contemporary Travel Writing*. Ann Arbor: University of Michigan Press.

Hopfenbeck, T. N. (2020). Making feedback effective? Assessment in education: principles. *Policy & Practice*, 27(1), 1–5. doi:10.1080/0969594X.2020.1728908

Hoskins W. (2003). *Devon*. Chichester: Phillimore.

Hsu, J. (2008). The secrets of storytelling: why we love a good yarn. *Scientific American Mind*, 18 September 2008. https://leyendomucho.blogspot.com/2008/09/secrets-of-storytelling.html, 20. 5. 2020.

Hudson, S., & Thal, K. (2013). The impact of social media on the consumer decision process: implications for tourism marketing. *Journal of Travel & Tourism Marketing*, 30(1/2), 156–160.

Huta, V., & Waterman, A. S. (2014). Eudaimonia and its distinction from hedonia: developing a classification and terminology for understanding conceptual and operational definitions. *Journal of Happiness Studies*, 15, 1425–1456.

Hutchins, Z. (2013). Travel writing, travel reading and the boundaries of genre: embracing the banal in Frankin's 1747 Pennsylvania Gazette. *Studies in Travel Writing*, 17(3), 300–319.

Ilić, J., Lukić, T., Besermenji, S., & Blešić, I. (2021). Creating a literary route through the city core: tourism product testing. *Journal of the Geographical Institute "Jovan Cvijić" SASA*, 71(1), 91–105.

INSEE (2016). *INSEE's Pack Hôtels Product*. Paris: INSEE.

Ittelson, W. (1978). Environmental perception and urban experience. *Environment and Behavior*, 10(2), 193–213.

Jasna, P. (2021). Montenegro Field Notes. Handwritten participant data. Unpublished.

Jenkins, H. (2006). *Convergence Culture. Where Old and New Media Collide New York*. London: New York University Press.

Jeremiah, M. (2000). The use of place in writing and literature. *Language Arts Journal of Michigan*, 16(2), 23–27.

Johnson, D. R. (2013). Transportation into literary fiction reduces prejudice against and increases empathy for Arab-Muslims. *SSOL*, 3(1), 77–92. doi:10.1075/ssol.3.1.08joh

Johnson, D. R. (2012). Transportation into a story increases empathy, prosocial behavior, and perceptual bias toward fearful expressions. *Personality and Individual Differences*, 52(2), 150–155.

Kafka, F. (1988). *The Diaries of Franz Kafka*. New York: Schocken Books.

Kaltenborn, B. (1998). Effects of sense of place on responses to environmental impacts: a study among residents in Svalbard in the Norwegian high Arctic. *Applied Geography*, 18(2), 169–189.

Kar, A. K., Kumar, S., & Ilavarasan, P. V. (2021). Modelling the service experience encounters using user-generated content: a text mining approach. *Global Journal of Flexible Systems Management*, 22(4), 267–288.

Kara, H. (2004). *How To Give Feedback on Academic Writing—Twelve TopTips*. December 2004. https://helenkara.com/2018/12/04/how-to-give-feedback-on-academic-writing-twelve-top-tips/, 2. 9. 2021.

Keskin, H., Akgun, A. E., Zehir, C., & Ayar, H. (2016). Tales of cities: city branding through storytelling. *Journal of Global Strategic Management*, 10(1), 31–41.

Kidd, D. C., & Castano, E. (2013). Reading literary fiction improves theory of mind. *Science*, 342, 377–380. doi:10.1126/science.1239918

Kim, M., & Kim, J. (2020). The influence of authenticity of online reviews on trust formation among travelers. *Journal of Travel Research*, 59(5), 763–776. doi:10.1177/0047287519868307

Kirillova, K., Fu X., Lehto X., & Cai, L. (2014). What makes a destination beautiful? Dimensions of tourist aesthetic judgement. *Tourism management*, 42, 282–293. https://www.emerald.com/insight/content/doi/10.1108/JHTT-02-2019-0041/full/html, 13. 6. 2022.

Kitsios, F., Kamariotou, M., Karanikolas, P., & Grigoroudis, E. (2021). Digital marketing platforms and customer satisfaction: identifying eWOM using big data and text mining. *Applied Sciences*, 11(17), 8032. https://www.mdpi.com/2076-3417/11/17/8032, 13. 6. 2022.

Kitsios, F., Mitsopoulou, E., Moustaka, E., & Kamariotou, M. (2022). User-generated content behavior and digital tourism services: a SEM-neural network model for information trust in social networking sites. *International Journal of Information Management Data Insights*, 2(1). doi:10.1016/j.jjimei.2021.100056; https://www.sciencedirect.com/science/article/pii/S2667096821000495, 13.6.2022.

Kneepkens, E., & Zwaan, R. (1994). Emotions and literary text comprehension. *Poetics*, 23, 125–138.

Knoop, C., Wagner, V., Jacobsen, T., & Menninghaus, W. (2016). Mapping the aesthetic space of literature "from below." *Poetics*, 56, 35–49.

Knudsen, D., & Greer, C. (2011). Tourism and nostalgia for the pastoral on the island of Fyn, Denmark. *Journal of Heritage Tourism*, 6(2), 87–98.

Knudsen, D., Metro-Roland, M., & Rickly, J. (2015). Tourism, aesthetics, and touristic judgment. *Tourism Review International*, 19(4), 179–191.

Koeck, R., & Warnaby, G. (2015). Digital Choreographies: conceptualising experiential representation and marketing of urban architectural geographies. *Architectural Research Quarterly*, 19(2), 183–192.

Koopman, E. (2016). Effects of "literariness" on emotions and on empathy and reflection after reading. *Psychology of Aesthetics Creativity and the Arts*, 10, 82–98.

Koopman, E. (2015). Empathic reactions after reading: the role of genre, personal factors and affective responses. *Poetics*, 50, 62–79.

Koščak, M. & O'Rourke, T. (2019). *Ethical and Responsible Tourism: Managing Sustainability in Local Tourism Destinations*. London: Routledge.

Kreziak, D., & Frochot, I. (2011). Co-construction de l'expérience touristique: les stratégies des tour-istes en stations de sport d'hiver [Co-construction of the tourism experience: tourists strategies in ski resorts]. *Décisions Marketing*, 64, 23–33.

Kristeva, J. (1980). *Desire in Language: A Semiotic Approaches to Art and Literature*. New York: Columbia University Press.

Kubacki, K. & Skinner, H. (2006). Poland: Exploring the relationship between national brand and national culture *Journal of Brand Management* 13(4), 284–299.

Kuijpers, M., & Hakemulder, F. (2018). Understanding and appreciating literary texts through rereading. *Discourse Processes*, 55(7), 619–641. doi:10.1080/01638 53X.2017.1390352

Kujanpää, M., Syrek, C., Lehr, D., Kinnunen, U., Reins, J. A., & de Bloom, J. (2021). Need satisfaction and optimal functioning at leisure and work: a longitudinal validation study of the DRAMMA model. *Journal of Happiness Studies*, 22, 681–707. doi:10.1007/s10902-020-00247-3, 15.4.2022.

Laing, H., & Frost, W. (2017). Journeys of well-being: women's travels narratives of transformation and self-discovery in Italy. *Tourism Management*, 62, 110–119.

Larsen, S. (2007). Aspects of a psychology of the tourist experience. *Scandinavian Journal of Hospitality and Tourism*, 7(1), 7–18. doi:10.1080/15022250701226014

Lee, K., & Ruck, K. (2022). Barista diary: an autoethnography studying the operational experience of third wave coffee shop baristas. *International Journal of Hospitality Management*, 102, 103–182.

Leung, X. Y., & Bai, B. (2013). How motivation, opportunity, and ability impact travelers' social media involvement and revisit intention. *Journal of Travel & Tourism Marketing*, 30(1/2), 58.

Lévinas, E. (1985). *Ethics and Infinity*. Pittsburgh: Duquesne University Press.

Li, C., Guo, S., Wang, C., & Zhang, J. (2019). Veni, vidi, vici: the impact of social media on virtual acculturation in tourism context. *Technological Forecasting and Social Change*, 145, 513–522. https://www.sciencedirect.com/science/article/abs/pii/S0040162517316104, 13. 6. 2022.

Lindqvist, S. (2018). *Exterminate all the Brutes*. London: Granta.

Liszewski, S., Bachvarov, M., (1998). Istota i właściwości prze-strzeni rekreacyjno-turystycznej, *Turyzm*, 8(1), 39–67.

Lucchesi, F. (1996). Geographic memories in travelogue literature: the Australian social landscape in the writings of Italian travelers. *GeoJournal*, 38(1), Geography and Literature (January 1996), 129–136 (8 pages), https://www.jstor.org/stable/41146710, 21. 1. 2021.

Lugmayr, A., Sutinen, E., & Suhonen, J. (2017). Serious storytelling – a first definition and review. *Multimedia Tools and Applications*, 76, 15707–15733. doi:10.1007/s11042-016-3865-5, 20.5.2020.

MacInnis, C., & Portelli, J. (2002). Dialogue as research. *Journal of Thought*, 37(2), 33–44.

Mailer, N. (1968). *The Armies of the Night: History as a Novel, the Novel as History*. New York: The New American Library.

Mailer, N. (1968). *Miami and the Siege of Chicago: An Informal History of the American Political Conventions of 1968*. Harmondsworth: Penguin Books.

Maitland, R., & Smith, A. (2009). Tourism and the aesthetics of the built environment. In J. Tribe (Ed.), *Philosophical Issues in Tourism* (pp. 171–190). Bristol: Channel View Publishing.

Mangold, W. G., & Faulds, D. J. (2009). Social media: the new hybrid element of the promotion mix. *Business Horizons*, 52(4), 357–365.

Mansfield, C. (2021). Way-tales: an archaeological topophonics for emerging tourist spaces. In B. Piga, D. Siret, & J. Thibaud (Eds.), *Experiential Walks for Urban Design. Revealing, Representing, and Activating the Sensory Environment* (pp. 323–332). London: Springer. Chapter 19. doi: 10.1007/978-3-030-76694-8_19

Mansfield, C. (2020). *Be Wallonia Wallonia Be*. Brussels: Wallonia Foreign Trade & Investment Agency. http://www.wallonia.be/en/blog/be-wallonia, 2. 2. 2021.

Mansfield, C. (2019). *The Role of Travel Writing Practitioners in Tourism Management and Place-branding Research*. ResearchGate.

McGaurr, L. (2012). The devil may care: Travel journalism, cosmopolitan concern, politics and the brand. *Journalism Practice*, 6(1), 42–58.

McKercher, B. (2016). Towards a taxonomy of tourism products. *Tourism Management*, 54, 196–208.

McKercher, B., Ho, P. S- Y., & du Cros, H. (2004). Attributes of popular cultural attractions in Hong Kong. *Annals of Tourism Research*, 31(2), 393–407.

McIntyre, C. (2009). Museum and art gallery experience space characteristics: an entertaining show or a contemplative bathe? *International Journal of Tourism Research*, 11(2) 155–170.

Mee, C. (2009). Journalism and travel writing: from grands reporters to global tourism. *Studies in Travel Writing*, 13 (4), 305–315. doi:10.1080/13645140903269072

Mehmetoglu, M., & Engen, M. (2011). Pine and Gilmore's concept of experience economy and its dimensions: an empirical examination in tourism. *Journal of Quality Assurance in Hospitality & Tourism*, 12(4), 237–255. doi:10.1080/1528008X.2011.54184

Miall, D. (1988). Affect and narrative: a model of response to stories. *Poetics*, 17(3), 259–272.

Miall, D. (1986). Emotion and the self: the context of remembering. *British Journal of Psychology*, 77, 389–397.

Miall, D., & Kuiken, D. (2002). A feeling for fiction: becoming what we behold. *Poetics*, 30, 221–241.

Miall, D., & Kuiken, D. (1998). Shifting perspectives: readers' feelings and literary response. In W. Van Peer & S. Chatman (Eds.), *Narrative Perspective: Cognition and Emotion* (SUNY Press, 2000). An earlier version of this paper was presented at the Conference on Narrative Perspective: cognition and Emotion. Zeist, The Netherlands, June 1–3 1995.

Michel, B. (2014). Les quartiers créatifs, entre clubbisation et ouverture du développement territorial. *ESO Espaces et Sociétés* (UMR6590), Travaux et documents, 37, 27–34.

Milton, J. (1671). *Paradise Regained*. London: John Starkey.

Mirrlees, H. (1920). *Paris: A Poem*. London: Hogarth Press.

Modiano, P. (2016). *Villa Triste*. London: Daunt.

Modiano, P. (1989). *Vestiaire de l'enfance*. Paris: Gallimard.

Monga, L. (1996). Travel and travel writing: an historical overview of hodoeporics. *Annali d'Italianistica*, 14, 6–54.

Morgan, M., Elbe, J., & Curiel, J. E. (2009). Has the experience economy arrived? The views of destination managers in three visitor-dependent areas. *International Journal of Tourism Research*, 11, 201–216.

Morris, B. (2004). What we talk about when we talk about "walking in the city". *Cultural Studies*, 18(5), 675–697.

Morris, J. (2002). *Trieste and the Meaning of Nowhere*. London: Faber.

Mossberg, L. (2007). A marketing approach to the tourist experience. *Scandinavian Journal of Hospitali-ty and Tourism*, 7(1), 59–74.

Mossberg, L., Therkelsen, A., Huijbens, E. H., Björk, P., & Olsson, A. K. (2010). *Storytelling and Destination Development*. Oslo: Nordic Innovation Centre.

Mumper, M. L., & Gerrig, R. J. (2017). Leisure reading and social cognition: a meta-analysis. *Psychology of Aesthetics, Creativity, and the Arts*, 11(1), 109–120. doi:10.1037/aca0000089, 13.1.2021.

Mundell, M. (2018). Crafting 'literary sense of place': the generative work of literary place-making. *Journal of the Association for the Study of Australian Literature*, 18, 1.

Myhill, D., Newman, R., & Watson, A. (2020). Thinking differently about grammar and metalinguistic understanding in writing. *Bellaterra Journal of Teaching & Learning Language & Literature*, 13(2), e870 pp. 1–19.

Nantes Tourisme (2015). *Le Voyage à Nantes*. Nantes.

Nawijn, J., Mitas, O., Lin, Y., & Kerstetter, D. (2013). How do we feel on vacation? A closer look at how emotions change over the course of a trip. *Journal of Travel Research*, 52(2), 265–274.

Nell, V. (1988). *Lost in a Book: The Psychology of Reading for Pleasure.* New Haven, CT: Yale University Press.

Newman, D. B., Tay, L., & Diener, E. (2014). Leisure and subjective well-being: a model of psychological mechanisms as mediating factors. *Journal of Happiness Studies*, 15(3), 555–578.

Nielsen, G. (2012). *The Norms of Answerability: Social Theory Between Bakhtin and Habermas.* New York: State University of New York Press.

Pabel, A., & Pearce, P. (2018). Selecting humour in tourism settings – a guide for tourism operators. *Tourism Management Perspectives*, 25, 64–70.

Pahor, B. (2018). *Place Oberdan à Trieste – Nouvelles.* Paris: Pierre-Guillaume de Roux.

Patron, S. (2020). *Small Stories – Une Nouveau Paradigm Pour les Recherches sur le Récit.* Paris: Hermann.

Patterson, M., & Williams, R. (2005). Maintaining research traditions on place: diversity of thought and scientific progress. *Journal of Environmental Psychology*, 25, 361–380.

Plank, C., Dixon, H., & Ward, G. (2014). Student voices about the role feedback plays in the enhancement of their learning. *Australian Journal of Teacher Education*, 39(9). doi:10.14221/ajte.2014v39n9.8

Porteous, J. (1996). *Environmental Aesthetics: Ideas, Politics and Planning.* London and New York: Routledge.

Potočnik Topler, J. (2022). The role of humour in tourism discourse: the case of Montenegro. *European Journal of Humour Research*, 10(1), 62–75. doi:10.7592/EJHR2022.10.1.639

Potočnik Topler, J. (2019). The role of new media in the development of tourism in Sevnica. In L. Zekanović-Korona (Ed.), *Informacijska tehnologija I mediji 2017: zbornik 2*, (Biblioteka Informacijska tehnologija i mediji, knj. 2) (pp. 71–78). Zadar: Sveučilište.

Potočnik Topler, J. (2018). Turning Travelogue readers into tourists: representations of tourism destinations through linguistic features. *Cuadernos De Turismo*, 42, 447–464. doi:10.6018/turismo.42.20

Potočnik Topler, J. (2017). Tourism discourse. In J. Potočnik Topler, A. Lisec, & M. Knežević (Eds.), *On Tourism Discourse and Other Issues in Tourism.* London: Pearson.

Potočnik Topler, J. (2016). *Literary Tourism: The Case of Norman Mailer, Mailer's Life and Legacy.* Frankfurt am Main: Peter Lang.

Potočnik Topler, J., & Zekanović-Korona, L. (2018). Digital media, perception and the selection of the 2016 Best European Destination: the case of Zadar. *Annales*, 28(2), 343–354. doi:10.19233/ASHS.2018.23

Prall, D. (1929). *Aesthetic Judgement.* New York: Crowell.

Proshansky, H. (1978). The city and self-identity. *Environment and Behaviour*, 10(2), 147–169.

Proshansky, H., Fabian, A., & Kaminoff, R. (1983). Place-identity: physical world socialization of the self. *Journal of Environmental Psychology*, 3(1), 57–83.

Proulx, M. (2003). *The Heart Is an Involuntary Muscle.* Vancouver: Greystone.

Proulx, M. (2002). *Le Cœur est un Muscle Involontaire.* Montréal: Boréal.

QAA (2019). *Creative Writing—Subject Benchmark Statement.* Gloucester: Quality Assurance Agency for Higher Education.

QAA (2014). *The Frameworks for Higher Education Qualifications of UK Degree-Awarding Bodies.* Gloucester: Quality Assurance Agency for Higher Education.

Rabinow, P. (2008). *Marking Time: On the Anthropology of the Contemporary.* Princeton and Oxford: Princeton University Press.

Rabinow, P. (2007). *Reflections on Fieldwork in Morocco*, Thirtieth Anniversary Edition. Berkeley: University of California Press.

Räikkönen, J., & Honkanen, A. (2013). Does satisfaction with package tours lead to successful vacation experiences? *Journal of Destination Marketing & Management*, 2(2), 108–117. doi:10.1016/j.jdmm.2013.03.00

Ram, Y., Björk, P., & Weidenfeld, A. (2016). Authenticity and place attachment of major visitor attractions. *Tourism Management*, 52, 110–122.

Ramaprasad, A. (1983). On the definition of feedback. *Behavioral Science*, 28(1), 4–13.

Reid, J. (1993). *Narration and Description in the French Realist Novel – The temporality of lying and forgetting.* Cambridge: Cambridge University Press.

Relph, E. (1976). *Place and Placelessness.* London: Pion.

Rice, J. (2015) Inscriptions of the possible. In S. Dobrin (Ed.), *Writing Posthumanism, Posthuman Writing (New Media Theory)* (pp. 155–173). Anderson: Parlor Press.

Richards, G., King, B., & Yeung, E. (2020). Experiencing culture in attractions, events and tour settings. *Tourism Management*, 79, 104–104.

Richins, M. (1997). Measuring emotions in the consumption experience. *Journal of Consumer Research*, 24 (September), 127–146.

Rihova, I., Buhalis, D., Moital, M., & Gouthro, M. B. (2014). Conceptualising customer-to-customer value co-creation in tourism. *International Journal of Tourism Research*. doi:10.1002/jtr.199

Robinson, M. (2004). Narratives of being elsewhere: tourism and travel writing. In A. A. Lew, M. C. Hall, & M. A.Williams (Eds.), *A Companion to Tourism* (pp. 303–316), Malden, Oxford, Carlton: Blackwell Publishing.

Robinson, M., & Andreson, H. C. (2002). *Literature and Tourism: Reading and Writing Tourism Texts.* London, Continuum.

Ross, K. (1988) *The Emergence of Social Space: Rimbaud and the Paris Commune.* Minneapolis and Markham: University of Minnesota and Fitzhenry & Whiteside Ltd, Canada.

Rötig, M. (2018). *Cargo.* Paris: Gallimard.

Roto, V., Jung-Joo L., Lai-Chong Law, E., and Zimmerman, J. (2021). The overlaps and boundaries between service design and user experience design. In *Designing Interactive Systems Conference 2021 (DIS '21), June 28-July 02, 2021, Virtual Event.* ACM, New York, NY, 13 Pages.

Rubiés, J., & Bacon, F. (2000). Travel writing as a genre. *Journeys*, 1(1), 5–35.

Russell, A. (2010) 'Isabelle de Montolieu Reads Anne Elliot's Mind: free Indirect Discourse in *La Famille Elliot. Persuasions*, 32, 232–247.

Sackville-West, V. (2007). *Passenger to Tehran.* London: Barbara Ward & Associates.

Sair, A., & Slilem, Y. (2022). Tourism destinations as Innovative clusters: an opportunity to make a destination more attractive: case of Agadir city. *International Journal of Accounting, Finance, Auditing, Management and Economics*, 3(4–1), 368–382.

Sardo, A., & Chaves, R. (2022). Literary tourism as a good practice to promote inland tourism: the case of the Eça de Queiroz Foundation in Portugal. In G. Fernandes (Ed.), *Challenges and New Opportunities for Tourism in Inland Territories: Ecocultural Resources and Sustainable Initiatives* (pp. 187–206). Hershey, PA: IGI Global.

Sartre, J. (2010). *The Imaginary.* London: Routledge.

Sartre, J. (1963). *Nausea.* London: Penguin.

Sartre, J. (1938). *La Nausée.* Paris: Gallimard.

Saunders, A., & Moles, K. (2016). Following or forging a way through the world: audio walks and the making of place. *Emotion, Space and Society,* 20, 68–74.

Scarles, C. (2004). Mediating landscapes: the processes and practices of image construction in tourist brochures of Scotland. *Tourist Studies,* 4, 43–67.

Scolari, C. (2009). Transmedia storytelling: implicit consumers, narrative worlds, and branding in contemporary media production. *International Journal of Communication,* 3, 586–606.

Sebald, W. (2006). *Campo Santo.* London: Penguin.

Sebald, W. (2002). *Vertigo.* London: Vintage.

Sebald, W. (1999). *Vertigo.* London: Penguin.

Séraphin, H. (2015). *Impacts of Travel Writing on Post-conflict and Post-Disaster Destinations: The Case of Haiti.* River Tourism: The Pedagogy and Practice of Place Writing, University of Plymouth, 3rd June, http://www.fabula.org/actualites/colloque-river-tourism-the-pedagogy-and-practice-of-place-writing_67346.php

Shamai, S. (1991). Sense of place: an empirical measurement. *Geoforum,* 22(3), 347–358.

Shamai, S., & Ilatov, Z. (2015). Measuring sense of place: methodological aspects. *Tijdschrift Voor Economische en Sociale Geografie,* 96(5), 467–476.

Shi, L., & Zhu, Q. (2018). Urban space and representation in literary study. *Open Journal of Social Sciences,* 6, 223–229.

Shor, I., & Freire, P. (1987). *A Pedagogy for Liberation.* Granby, MA: Bergin & Garvey.

Shute, V. (2008). Focus on formative feedback. *Review of Educational Research,* 78(1), 153–189.

Sikora, S., Kuiken, D., & Miall, D. S. (2010). An uncommon resonance: the influence of loss on expressive reading. *Empirical Studies of the Arts,* 28(2), 135–153.

Singer, R. (2016). Leisure, refuge and solidarity: messages in visitors' books as microforms of travel writing, *Studies in Travel Writing,* 20(4), 392–408. doi:10.1080/13645145.2016.1259606

Škulj, J. (2004). Literature and space: textual, artistic and cultural spaces of transgressiveness. *Primerjalna Književnost,* 27, 21–37.

Slovenia (2017). *Strategy for the Sustainable Growth of Slovenian Tourism for 2017–2021, Slovenia Ministry of Economic Development.* https://www.slovenia.info/uploads/publikacije/the_2017-2021_strategy_for_the_sustainable_growth_of_slovenian_tourism_eng_web.pdf, 21. 6. 2022.

Smaldone, D., Harris, C., Sanyal, N., & Lind, D. (2005). Place attachment and management of critical park issues in Grand Teton National Park. *Journal of Park & Recreation Administration,* 23, 1.

Smith, S. (2015). A sense of place: place, culture and tourism. *Tourism Recreation Research,* 40(2), 220–233.

Smith, Y. (2012). *Literary Tourism as a Developing Genre: South Africa's Potential.* Pretoria: University of Pretoria (dissertation).

Sporre, J. (2006). *Perceiving the Arts: An Introduction to the Humanities,* Eighth Edition. Upper Saddle River, NJ: Pearson Prentice Hall.

Stedman, R. C. (2003). Is it really just a social construction? The contribution of the physical environment to sense of place. *Society & Natural Resources,* 16(8), 671–685.

Stedman, R., Beckley, T., Wallace, S., & Ambard, M. (2004). A picture and 1000 words: using resident-employed photography to understand attachment to high amenity places. *Journal of Leisure Research*, 36(4), 580–606.

Stepchenkova, S., & Zhan, F. (2013). Visual destination images of Peru: comparative content analysis of DMO and user-generated photography. *Tourism Management*, 36, 590–601.

Stiebel, L. (2007). Hitting the hotspots: literary tourism as a research field in KwaZulu-Natal. *Critical Arts*, 18(2), 2004.

Stracke, E., & Kumar, V. (2020). Encouraging dialogue in doctoral supervision: the development of the feedback expectation tool. *International Journal of Doctoral Studies*, 15, 265–284.

Stuart Chapin, F., & Knapp, C. (2015). Sense of place: a process for identifying and negotiating potentially contested visions of sustainability. *Environmental Science & Policy*, 53, Part A, 38–46.

Sundin, A., Andersson, K., & Watt, R. (2018). Rethinking communication: integrating storytelling for increased stakeholder engagement in environmental evidence synthesis. *Environmental Evidence*, 7(6), doi:10.1186/s13750-018-0116-4

Tamboukou, M. (1999). Writing genealogies: an exploration of Foucault's strategies for doing research. *Discourse: Studies in the Cultural Politics of Education*, 20(2), 201–217.

Tan, S.-K., Kung, S.-F., & Luh, D.-B. (2013). A model of 'creative experience' in creative tourism. *Annals of Tourism Research*, 41, 153–174. doi:10.1016/j.annals.2012.12.002

Thexton, T., Prasad, A., & Mills, A. (2019). Learning empathy through literature. *Culture and Organization*, 25(2), 83–90.

Thompson, C. (2011). *Travel Writing*. New York: Routledge.

Tiede, M. (2016). *The Importance of Travel Journalism*. Fort Worth, TX: Texas Christian University (dissertation).

Tierno, M. (2002). *Aristotle's Poetics for Screenwriters: Storytelling Secrets from the Greatest Mind in Western Civilization*. New York: Hyperion.

Todd, C. (2012). The importance of the aesthetic. In A. Holden & D. A. Fennel (Eds.), *The Routledge Handbook of Tourism and the Environment* (pp. 65–74). London and New York: Routledge.

Tuan, Y. (1991). Language and the making of place: a narrative-descriptive approach. *Annals of the Association of American Geographers*, 81(4), 684–696.

Tuan, Y. F. (1977). *Space and Place*. London: Edward Arnold.

Tuan, Y. F. (1974). *Topophilia: A Study of Environmental Perception, Attitudes, and Values*. Englewood Cliffs, NJ: Prentice Hall.

Tung, V., & Ritchie, J. (2011). Exploring the essence of memorable tourism experiences. *Annals of Tourism Research*, 38(4), 1367–1386.

Tursić, M. (2019). The city as an aesthetic space. *City*, 23(2), 205–221.

Van Assche, K., Beunen, R., & Oliveira, E. (2020). Spatial planning and place branding: rethinking relations and synergies. *European Planning Studies*, 28(7), 1274–1290.

Van Laer, T., Feierisen, S., & Visconti, L. M. (2019). Storytelling in the digital era: a meta-analysis of relevant moderators of the narrative transportation effect. *Journal of Business Research*, 96, 135–146.

Vitić-Ćetković, A., Jovanović, I., & Potočnik Topler, J. (2020). Literary tourism: the role of Russian 19th century travel literature in the positioning of the smallest

European Royal Capital – Cetinje. *Annales, Series Historia et Sociologia*, 30(1), 81–99.

Vorkinn, M., & Riese, H. (2001). Environmental concern in a local context: the significance of place attachment. *Environment and Behavior*, 33(2), 249–263.

Wagoner, B., Brescó, I., & Awad, S. (2019). *Theories of Constructive Remembering*. London: Springer.

Weber, D., Wroge, D., & Yoder, J. (2007). Writers' workshops: a strategy for developing indigenous writers. *Language Documentation and Conservation*, 1(1), 77–93.

Więckowski, M. (2014). Tourism space: an attempt at a fresh look. *Turyzm*, 24(1), 17–24.

Wilson, A., Murphy, H., & Cambra Fierro, J. (2012). Hospitality and travel: the nature and implications of user-generated content. *Cornell Hospitality Quarterly*, 53(3), 220–228.

Woodside, G., & Megehee, C.M. (2010). Advancing consumer behaviour theory in tourism via visual narrative art. *International Journal of Tourism Research*, 12(5), 418–431.

Writing-center (2020). *News Writing Fundamentals*. George Mason University. https://writingcenter.gmu.edu/guides/news-writing-fundamentals, 27. 5. 2022.

Xiang, Z., & Gretzel, U. (2010). Role of social media in online travel information search. *Tourism Management*, 31, 179–188.

Xu, H., Cheung, L.T.O., Lovett, J., Duan, X., Pei, Q., & Liang, D. (2021). Understanding the influence of user-generated content on tourist loyalty behavior in a cultural World Heritage Site. *Tourism Recreation Research*. doi:10.1080/025082 81.2021.1913022, https://www.tandfonline.com/doi/full/10.1080/02508281.2021.19 13022, 13.6.2022.

Yousaf, S. (2022). Food vloggers and parasocial interactions: a comparative study of local and foreign food vlogs using the S-O-R paradigm. *International Journal of Contemporary Hospitality Management*. doi:10.1108/IJCHM-09-2021-1090

Yuan, Y., Major-Girardin, J., & Brown, S. (2018). Storytelling is intrinsically mentalistic: a functional magnetic resonance imaging study of narrative production across modalities. *Journal of Cognitive Neuroscience*, 30(9), 1298. doi:10.1162/jocn_a_01294

Yuksek, A., Yuksel, F., & Bilim, Y. (2010). Destination attachment: effects on customer satisfaction and cognitive, affective and conative loyalty. *Tourism Management*, 31(2), 274–284.

Zacharia, C. (2021). Slovenia's new Juliana Trail reveals a land of water, rock and forest. *The Guardian*. https://www.theguardian.com/travel/2021/aug/11/slovenia-juliana-walking-trail-sustainable-tourism-national-park, 23. 2. 2022.

Zakarevičius, P., & Lionikaitė, J. (2013). An initial framework for understanding the concept of internal place branding. *Management of Organizations: Systematic Research*, 67, 143–160.

Zeng, B., & Gerritsen, R. (2014). What do we know about social media in tourism? A review. *Tourism Management Perspectives*, 10(1), 27–36.

Zenker, S., & Erfgen, C. (2014). Let them do the work: a participatory place branding approach. *Journal of Place Management and Development*, 7(3), 225–234.

Zhang, Z., Zhang, Z., & Law, R. (2014). Positive and negative word of mouth about restaurants: exploring the asymmetric impact of the performance of attributes. *Asia Pacific Journal of Tourism*, 19(2), 162–180.

Zilcosky, J. (2008). *Writing Travel: The Poetics and Politics of the Modern Journey*. Toronto: University of Toronto.

Zube, E., Sell, J., & Taylor, G. (1982). Landscape perception: research, application and theory. *Landscape Planning* 9, 1–33.

# INDEX